PRAISE FOR ONE IS A WARRIOR

Harry Martinez is one of those rare Soldiers who runs toward the sound of the guns and stays there when others falter. I have witnessed his steadiness under pressure and his loyalty to the men around him. What sets him apart is not simply that he fought—it is that he refuses to lie about what fighting does to a man. For those who have served, you will recognize the truth in these pages immediately. ~Lieutenant Colonel Nathan Wall, Graduate of War College and Yale and Harvard

Harry Martinez writes with the clarity of a man who understands not only tactics, but consequence. He is a warrior in the truest sense. He is disciplined, capable, and resolute. What sets this work apart is his willingness to confront the moral weight that follows the trigger pull. That awareness does not weaken the warrior; it refines him.

ONE IS A WARRIOR is not for those seeking entertainment. It is for those willing to understand the burden carried long after the firefight ends. Harry does not ask for sympathy. He offers honesty. And that honesty is rare. ~Dustin Shepherd, Retired US Army Sergeant Major, Professor of Executive Leadership

I was in Ramadi during the account in ONE IS A WARRIOR. When our unit came under sniper attack from opposition forces, I was overcome by the sense of comfort—and even a bit of confidence—when I heard over the radio, "Shadow 4 is your overwatch." ~Richard Hughes

This book will take you on a harrowing journey through one of the most dangerous theaters of war on the planet. It is not filtered through a Hollywood lens. What this work offers is something far more valuable: an unvarnished immersion into the warrior's process. It is a journey defined by experimentation, creativity under fire, hard-learned mistakes, moral uncertainty, self-doubt, and the spiritual wounds that are the true cost of combat — wounds that rarely make the headlines, but that every warrior carries long after the mission ends. ~Kevin Mulcahey, Command Sergeant Major (Retired) 1/172nd Cavalry (Mountain)

ONE IS A WARRIOR is a visceral, unfiltered portrayal of Ramadi as we experienced it—raw, immediate, and unflinching with split-second decisions amidst the chaos of urban combat. He captures the sensory overload of battle, the haunting silence after a shot, and the heavy toll that lingers long after the adrenaline fades, and the pervasive guilt that follows Soldiers home. Semper Fi. ~Jason Angell, Author of *Running Towards Gunfire: Courage and Brotherhood in Ramadi,* USMC Veteran

ONE IS A WARRIOR opens with calculated precision—night insertions by helicopter, coordinated convoy movements, and small units slipping into position under cover of darkness. From the outset, the narrative establishes credibility and immediacy, placing the reader alongside a sniper team infiltrating Ramadi. What follows is a disciplined account of combat, where engagements are executed with clarity, control, and purpose.

This is a **direct, unembellished account of modern combat**—one that captures both the operational reality and the human dimension with clarity and authority. ~(RET) First Sergeant Richard Hughes, ARNG, Marine Veteran

ONE IS A WARRIOR is a gripping and insightful transformation of a humble soldier into a battle-hardened Warrior forged by the ravages of modern warfare. This close and personal account places you into the trials and hardships that force strong men to rise above the uncertain, chaotic, bloody, street-to-street battles using intellect, skill, and irrefutable bravery. This work will engage you in the victories the author earned through perseverance, personal struggles, hardship, and an unwavering dedication to mission accomplishment. **These riveting experiences demonstrate the evolution and adaptability of snipers, making these teams the most lethal asset in a commander's arsenal.** ~Jacek Niemynski, (RET) Command Sergeant Major, IN New Jersey Army National Guard

In ONE IS A WARRIOR, the author presents a raw and unfiltered account of combat experience that resonates deeply with anyone familiar with the realities of warfare. The battle scenarios are depicted with striking clarity, illustrating not only the physical challenges but also the mental fortitude required to thrive in such environments. **For anyone seeking an authentic portrayal of military life and the sacrifices made in service, ONE IS A WARRIOR is an essential read.** ~Vincent Caliguire, Army Veteran

In ONE IS A WARRIOR, SGM Harry Martinez shares over two decades of service across diverse threat environments. He draws upon a wealth of experience that resonates deeply with both military personnel and civilians alike. Harry masterfully articulates the sacrifices endured not only by soldiers but also by their families, highlighting the emotional toll that comes with being at the forefront of combat. He emphasizes that the act of pulling the trigger is far more than a simple motion; it encompasses the weight of responsibility and the complexities of leadership.

For anyone seeking a profound understanding of the warrior ethos, ONE IS A WARRIOR is a must-read. It's a testament to the spirit of resilience and sacrifice that defines our armed forces. ~T. S. Decker, (RET) Command Sergeant Major, IN ARNG, Marine Veteran

"Peer through the narrow tunnel of s scope, and bear the life long weight of decisions measured in heartbeats in One is a Warrior."

—JASON ANGELL

Author of Running Towards Gunfire

ONE IS A WARRIOR

The Relentless Fight of an Army National Guard Sniper Team in Ramadi

SGM HARRY R. MARTINEZ (RETIRED)

Printed in the United States of America.

Paperback Edition ISBN: 978-1-972059-00-5
Hardcover Edition ISBN: 978-1-972059-01-2
Digital Edition ISBN: 978-1-972059-02-9
Large Print Edition ISBN: 978-1-972059-03-6
Library of Congress Control Number: 2026907080

Cover design by r2cdesign

Cover photograph ©Harry R. Martinez,
Harry Martinez in Ramadi, Iraq

Vendela Publishing, LLC, is registered in the
United States of America and other countries.

Out of every hundred men, ten should not be here,
eighty are just targets, nine are the real fighters,
and we are lucky to have them, for they make the battle.
Ah, but the one, one is a warrior, and
he will bring the others back.

DEDICATION

For my children, and The Bastards of 171, The Grunts: USMC 1/6, Army 2/69. 1/506, TF Dark Eagle, 10th SFG ODA 0311, CHOSEN 1-185 IN, 79th IBCT S3 Staff, and The Reason Outdoors who stood beside me through loss, who reminded me who I was when I forgot, and who never asked me to become smaller to survive.

TABLE OF CONTENTS

FOREWARD

ONE IS A WARRIOR is not a polished, arm's-length recounting of the Iraq War. It is Ramadi as we lived it, raw, immediate, and unflinching. Harry Martinez pulls no punches in describing the weight of split-second decisions, the sensory overload of urban combat in a city reduced to rubble, the moral razor's edge of the rules of engagement, and the quiet aftermath when adrenaline fades and the human cost settles in. In his vivid, unfiltered style, Harry brings the reader inside the green tunnel of night vision, the view from a seasoned sniper's scope, the heavy silence after a shot, and the lingering echoes of lives taken or spared.

I first crossed paths with Harry Martinez in the dust and darkness of eastern Ramadi during the brutal months of early 2006. As a team leader with the United States Marine Corps' 1st Air Naval Gunfire Liaison Company (ANGLICO), my small team of Marines and I were tasked to insert into the shadows and crevices of that grey dismal city, identify targets for close air support and direct the airstrikes in support of U.S. Army units locked in relentless combat. Harry's sniper team, *Joker 4*, was one of the teams we supported. We worked side by side as part of *Task Force Dark Eagle*, and Harry's team stood out from the rest like wolves among dogs. In that environment, trust is earned the hard way. I trusted him with my life and still do. His impact was so significant that he was a major part of my own book, *Running Towards Gunfire: Courage and Brotherhood in Ramadi.*

ONE IS A WARRIOR captures what so many accounts gloss over when writing about combat in Ramadi, the bone-deep exhaustion, the brotherhood forged in shared misery, the guilt that follows you home, the knowledge that the fight doesn't end when the deployment does. Harry writes honestly about the transformation war demands, how regular soldiers become hunters of men, how training and instinct take over in the moment, and how the faces return in the quiet hours long after the shooting and explosions stop. ONE IS A WARRIOR recounts the constant and frustrating battles that take place between soldiers and their egos, leadership, military bureaucracy and others that increase the self-imposed friction and chaos of war. Harry masterfully describes how they were forced to deal with this and with the enemy.

The title draws from a well-known quote:

> Out of every hundred men, ten should not be here,
> eighty are just targets, nine are the real fighters,
> and we are lucky to have them, for they make the battle.
> Ah, but the one, one is a warrior,
> and he will bring the others back.

Harry was that warrior for his men. He took a group of young soldiers, forged them into a lethal sniper team, led them through the deadliest months in Ramadi, and brought them home. These soldiers, and many others, reunited with their families because Harry pushed them to be aggressive fighters that ignored the fear of death. This gave them an edge. More importantly, the book describes the unvarnished truth of what it cost – and the cost was high.

If you want to study proven tactics, techniques, and procedures employed by an extremely effective sniper team, ONE IS A WARRIOR is essential reading. If you're looking for a clean, heroic gloss-over of the war, this isn't that book. But if you want to know what it actually felt like to move through those streets, to live under constant threat, to peer through the narrow tunnel of a scope, to bear the lifelong weight of decisions measured in heartbeats—then ONE IS A WARRIOR is for you. Share it with those who ask what combat really costs. Pass it to the next generation, one day they may step into the crucible of combat.

Semper Fi.

Jason Angell
Author of *Running Towards Gunfire: Courage and Brotherhood in Ramadi.*
USMC Veteran

ACKNOWLEDGEMENTS

No one comes through war alone, even when they walk it in silence. This book exists because of the people who stood beside me, both during the war and in the years that followed, when the noise faded, but the weight remained.

I owe my survival first to the people I served with—the ones who shared fear, exhaustion, and the kind of dark humor that appears when we had nothing left to hold onto. In moments when trust was the only currency that mattered, you showed me what it meant to rely on one another completely. What we endured together shaped this story, even when it resisted being put into words.

To those who did not make it home, and to those whose battles continued long after the fighting stopped—I carry your names with me. This book is written in your shadow, with the hope that it reflects something true about the cost of service and the lives that were altered or lost along the way.

My family lived with the long aftermath of war: the pauses, the distance, the things I could not explain and did not yet understand myself. You accepted the silences, endured the changes, and stayed when leaving would have been easier. Your presence made survival possible. Thank you, Joshua, James, and Jessica. I love you with every fiber of my core

I am also grateful to the counselors and the Brotherhood who helped me find a way forward when I did not know how to return to a quieter world. Your patience, honesty, and refusal to rush the process were a form of mercy.

This memoir does not offer clean answers or easy resolutions. It is an attempt to reckon with what remains. This story is about resilience—and the cost it demands.

INTRODUCTION

CPL Martinez assigned to STA 1/6. _Camp Lejeune spring of 1991

When I arrived in Ramadi in June 2005, I was a 38-year-old veteran police officer from a small town in New Jersey—a National

Guardsman and former Infantry Marine. My team, originally known as Shadow 4, consisted of me and three college students who had joined the Army National Guard for tuition assistance, a shot at an education and a better future. We ended up in Ramadi, the capital of Al Anbar Province—a searing hellscape at the heart of the "Sunni Triangle of Death" to fight through one of the bloodiest chapters of Operation Iraqi Freedom II and III.

Initially, my Soldiers trained as tank operators. When we all volunteered to become snipers, we entered a gauntlet of violence, exhaustion, and moral ambiguity that redefined who we were.

Reassigned to fight alongside the 3rd Infantry Division 2/69 Armor HHC SNIPER SECTION and later embedded with the 101st ABN DIV 1/506 IN Recon Platoon, we became Shadow 4 (later Joker 4)—a sniper team with a singular mission: to locate, observe, and kill the enemy.

Often operating at distances beyond 700 meters, we left a bloody footprint across the city, yet some missions brought us so close to our targets that we could see the fear in their eyes before the shot landed. Every moment was a battle—both with the enemy and within ourselves. Sniping was a relentless test of endurance, precision, and morality.

A city of 400,000, Ramadi was a defining trial in the Global War on Terrorism. Every street corner was a potential battlefield, and the lines between friend and foe blurred among its citizens. For the men and women who fought there, Ramadi symbolized the relentless grind of the Iraq War. That year, it earned its reputation as the most dangerous city on earth, the brutal reality faced by both the U.S. military and the civilian population.

In this city, chaos was the norm, and survival was a precarious daily battle. Ramadi drew the keen interest of the United States Armed Forces for several compelling reasons: strategic importance, heavy insurgency activity, and relentless violence.

Perched on critical routes connecting Baghdad to the west, Ramadi was the lifeblood of western Iraq. For the insurgents, it was a crown jewel. Seizing it allowed them to choke supply lines and assert power across the region. For U.S. forces, losing Ramadi would have shattered the already-unstable situation in Iraq, turning a conflict into outright chaos. Every street, every corner, became a battlefield where the stakes were nothing less than victory or failure.

When we arrived, Ramadi was a war zone. Al-Qaeda in Iraq entrenched itself here, blending their foreign fighters with disenfranchised locals and embittered Ba'ath Party loyalists. This unholy alliance weaponized the resentment of Ramadi's population. The city was consumed by violence. Many of its citizens co-opted into the insurgents' cause, but many did not. Innocent families were caught in the middle. Ramadi was the heart of a war machine, a cause for both sides.

For U.S. forces, Ramadi was a waking nightmare. The streets were dangerous. Ambushes, sniper fire, and the constant threat of IEDs turned every patrol into a deadly gamble. Each step could trigger an explosion, and every unseen dark corner concealed an enemy. Death waited at the end of every breath. Soldiers stayed on edge, sleep-deprived, and battle-hardened. The violence was relentless, where the only option was to maintain the will to fight under the umbrella of confidence.

Insurgents found the perfect hunting ground in Ramadi's urban sprawl. Narrow alleyways, concrete homes converted into bunkers, and industrial zones provided endless hiding places and kill zones. Combatants hid among civilians, slipping into crowds before re-emerging to strike. The urban landscape leveled the playing field for them against the superior force of U.S. troops. Ramadi was a labyrinth of death.

The price of this conflict was staggering. U.S. military losses mounted, and the toll on the civilian population was just as devastating. Innocent families lived in constant fear, caught in the crossfire between insurgents and Coalition Forces. The city was drenched in grief, distrust, and despair.

Every day was a battle of the mind for the Soldiers and Marines stationed in Ramadi. Under the constant threat of death, survival often came down to split-second decisions, and the weight of those choices took a heavy psychological toll long after the battles ended.

This story is a reminder of the sacrifices in that dark chapter of history, both by the Soldiers and Marines as well as the civilians who endured unimaginable hardship.

It is a raw, unflinching look at the savagery and complex nature of modern urban combat. Armed with nothing but radios, rifles, and grenades, we stepped into a city of 400,000 people, knowing that our sole purpose was to eliminate an enemy who may never know we were there until too late.

In these pages, you will find yourself in the sniper's hide, immersed in the harrowing world of Shadow 4. You'll feel the suffocating tension as endless hours stretch into days, eyes fixed on a scope, waiting for a target to appear. You'll come face-to-face with the quiet terror that chews away at your sanity because every moment could be your last. Our comrades fell, and our only choice was to suppress our grief and press on, mission after mission.

What lies within these pages is brutal, visceral, and savage. This is war.

ONE IS A WARRIOR

CHAPTER 1

FEET ON THE GROUND

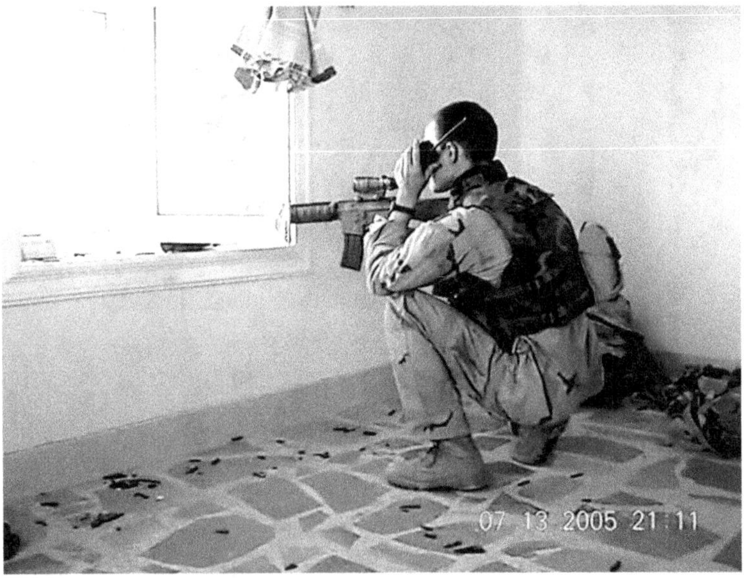

Ambushing an IED cell attempting emplace the deadly device, notice the shell casings on the floor, Ramadi Iraq July 2005

Pre-dawn darkness enveloped us as we crossed the road, our boots lightly crunching over the gritty earth. Ahead lay Combat Outpost, which we called simply The COP, home to an infantry unit hardened by their time in Ramadi.

Linking up with another squad, we moved in single file, making our way toward yet another gate. A hulking, five-ton truck loomed nearby. Next to it, a Soldier stood watch at the gate, looking every bit like a sentry at a medieval booth. His eyes scanned each of us as we filed past. We adjusted our gear and NVGs (night vision goggles), preparing ourselves for whatever waited beyond.

A last-minute headcount wafted through the line. I turned to my right, focusing on the Soldier in front of me. The gate groaned and slid open, revealing the city just beyond. One by one, the patrol leader counted each man as we stepped out, our presence dissolving into the murky, hostile expanse of Ramadi. Before I moved through, I glanced over my shoulder at the Soldier guarding the gate. Our eyes met—a moment of shared understanding, of shared dread. His face betrayed a flicker of fear. With a look of gut-wrenching urgency, he slammed the gate shut, the metallic clang echoing behind me.

Damn. If he's that scared, what the hell am I walking into?

I was the Shadow 4 Team Leader, and I'd best start acting like it. Stepping into the open, my field of vision was narrow and tinted green through the lens of my NVGs.

During the war from 2005 to 2006, the landscape of Ramadi, Iraq, showed the marks of destruction caused by months of open conflict. The city was a ghost town, its silence unnatural, its emptiness unsettling. The streets appeared lifeless. Buildings stood dark and hollow. Piles of garbage lined the sidewalks. The stench of decay clung to the thick air, a reminder of the city's deadlocked struggle with insurgents, organized civilian groups who fight for political change against the country's government.

After U.S. coalition forces invaded Iraq in 2003, a fierce Sunni insurgency rose against both the coalition and the new Shia-led government. The most dangerous among them was Al-Qaeda in Iraq (AQI), and Ramadi was a major AQI stronghold. Instead of a cohesive, organized army, they were a loose alliance of foreign jihadists, local fighters, and anyone else who wanted to join them. Some of them were foreign fighters who planned major attacks like suicide bombings, while others were locals with an ax to grind. Taking advantage of the chaos, criminal groups carried out smuggling, weapons trafficking, and kidnappings. When I arrived, AQI controlled some areas of Ramadi—running their own courts, killing anyone they suspected of helping the U.S., and harassing the coalition with constant attacks.

Buildings were often reduced to rubble, with many structures damaged or destroyed. Debris, chunks of broken concrete, twisted metal, and shattered vehicles filled the streets, making them almost impassable. Bullet casings lay everywhere. Pavement was pockmarked from IED blasts, each marking a skirmish in a never-ending battle. Military vehicles, checkpoints, and barricades were commonplace.

Dust and dirt coated everything in this desert land. Burned vehicles, damaged infrastructure, and signs of violence dotted the landscape. Gunfire and explosions echoed through the city. Amid this chaos, local communities tried to carry on with their lives. The once-vibrant community was now desolate, with many residents displaced or hiding in fear. Every step was cautious. Each corner held a potential threat.

Buildings along the streets were hollow shells, their windows shattered. Bullet holes and blackened scars marred their walls. Markets and shops were boarded up or looted. Occasionally, someone scrawled Arabic words across the walls.

The smell of burning refuse mingled with a deeper, more ominous odor—gunpowder and decay. Distant gunfire, the heavy rumble of military vehicles, and the sudden crack of a distant explosion made the city feel eerie, stripped of its soul, where every step forward could be your last—for soldiers and civilians alike. That night, I became part of the landscape.

As we advanced, we fell into a rhythm. At each intersection, a Soldier would pause and scan for movement, weapon ready. The Soldier behind him would tap his shoulder, signaling him to move on. I noticed the silent language and mimicked their motions. The pulse of their battle-worn discipline seeped into me. We had no room for mistakes—only precision, focus, and resolve. We were here to fight. We were here to win.

Halfway through town, the Soldier before me shot me a glance that said, *You are too fucking close.* I backed off.

We reached a large crossing and broke into a run. My heavy Night Vision Goggles caused the front of my helmet to slide down. I hadn't properly tightened my helmet. It was an issue but now was not the time to fix it. I could have kicked myself for such stupidity. I should have known better.

As we ran hard and fast across this wide street, my helmet slid down to the bridge of my nose. When I reached the other side, my head was leaning back, and I was peeking out beneath my helmet. The edge of the helmet cut into my skin. I kept my mouth shut.

As we reached our destination, I was impressed with the squad's navigation ability. We gained entry and began to clear the building. I was assigned to work with SPC AA, a young man with blond hair that was long on top and cropped close on the sides. He had a quick and confident smile.

After clearing the building without incident, AA dictated our next action. We took furniture from other rooms and built a shooting platform. Opening all the windows, we moved around to find places for well-placed shots.

AA told me to rest because we would probably be here all day. I told him I was too ramped up to sleep. If it was all the same to him, I would stay awake. The infantry squad took up positions throughout the building to ensure 360 security.

As the sun rose, the city came to life. Donkeys pulled carts. Motorcycles roared down the street. Vendors opened their shops. Across the way, an amusement park stood empty, as though waiting for better days. For a moment, I almost forgot there was a war going on. It seemed bizarre that an American squad lay in ambush, ready to kill on a busy city street in Iraq. With the dawn, my awareness also woke up. My entire body was as tight as a piano wire, my senses screeching.

The sun splashed its rays across our sector of the city. AA and I discussed possible trends. We focused on a white four-door sedan that drove by our position several times. At one point, a small crowd of young men appeared on the northeast corner. A young man was cleaning a farmer's tractor, but his attention was focused elsewhere. Periodically, he would glance to his left and right. His subtle movements were a reminder of quiet tension simmering beneath the surface.

As we discussed the situation, AA's ICOM radio squawked. At that moment, I was focused on the west side of the street. A jarring sound. I scanned my field of vision. Positioned to my left and facing north, AA focused on his side of the street, but he had a blind spot between his view and the corner. That small patch of "dead space" was a detail that could go unnoticed until it was too late. In Ramadi, you had to be aware of more than what you could see. What you couldn't see could be far more deadly.

A short, stocky Soldier with a stoic bearing, Shadow 1's team leader, Sergeant SH called out, "IED team inserting a device near the northeast corner!"

As I leaned forward, a single shot cracked through the air. A man lay in the street, slumped over on his right side, his back to us, utterly still. Before the echo could settle, a blue van barreled around the corner, screeching to a halt next to the fallen insurgent. Bold and foolhardy, they tried to pull their dead out of the noisy street. The growl of an

M240 machine gun tore through their efforts, unleashing a stream of rounds that sent them scrambling.

In an instant, insurgents flooded into the street from every direction. Aks and pistols glinting in their hands, they moved with a jerky kind of panic. Their confusion and fear seemed out of place with the firepower they held.

My heart pounded as I adjusted my grip on my rifle, instinct kicking in. *Breathe, relax, aim, shoot.* My sights locked on target after target, my reflexes moving faster than conscious thought. I raised my rifle and shot a man running toward the low wall. He almost collapsed, clutching his stomach as his weapon slipped from his grip. He staggered, screaming, pain contorting his face, while his comrades scrambled to drag him behind cover. They were so close I could make out the details of their faces, the expressions of shock and desperation as they hurried to shield him.

In that strange, surreal moment, I noticed the stubble on their jaws from hurried, uneven shaves. Some of them had clearly rushed through their morning routines. Others had taken time. It struck me as an oddly human detail, a glimpse into lives interrupted and thrown into inhuman circumstances.

I resumed scanning my sector, my eyes sharp, my ears attuned to any sound or movement that might signal danger. Suddenly, a shout rang out—followed by a straightforward command that rang through the room: "If anyone goes near the IED, kill them," and we knew one misstep could mean death.

A tan four-door car rolled southward toward our position, slowing as it neared the disabled van. The passenger stepped out and dashed toward the van as if to lend a hand. In a normal setting, this could be an act of kindness, a stranger stopping to help, but not here.

To Marines and Soldiers who had seen too much to believe in chance or mercy, he was an enemy disguised as a good Samaritan. In the cold calculus of survival, hesitation was fatal. Within seconds, he was struck down by a hail of bullets, his body crumpling in the dust.

I turned my sights to the car itself, lined up my sights, breathing steadily as I placed round after round into the driver's side, each shot landing precisely in a tight circle. The windshield fractured with spiderweb cracks, glass splintering inwards as the bullets hammered through. I continued, targeting the engine compartment to disable the vehicle for good. I had no second thoughts or regrets, just the mindless need to end the threat.

We had no time to think of the lives behind the faces, the stories behind the bodies. It was kill or be killed, the ugly arithmetic of war. It shaped every move we made, every trigger we pulled.

A passenger next to the driver climbed out. I shot him. He fell into a sitting position against a pole, tried to get up but failed. I thought of killing him but decided not to. He wasn't going anywhere, so I left him alone. My first critical mistake that day.

Around the same time, a man crouched behind three young boys, walking from north to south toward my position. He had the youngsters in a half circle in front of him like a human shield.

He approached the tan four-door sedan and popped the trunk. He tried to bring something out. An American voice yelled out his actions. The man kept the boys near him the whole time. Watching him, I slowed my breath and found a rhythm. I squeezed the slack out of my trigger. My weapon recoiled slightly.

He grabbed his left side and slammed to the ground. His human shields ran to my left, finding a corner for cover. Small hands reached out to him as he tried to drag himself towards safety. His hand extended, beckoning the boy to come closer.

I shot him repeatedly. The first few shots hit him in the neck. His head hit the pavement face-first and remained that way. A Soldier's voice reported a man traveling southbound along the far sidewalk. A Non-Commissioned Officer (NCO) ran inside and said, "If he tries to get near that vehicle, shoot him!"

The man wore a white short-sleeved shirt, tan trousers cinched by a dark belt, and polished dark shoes that seemed out of place in this chaos. He was slightly overweight, his belly pressing against his waistband. His bald head gleamed with sweat, intense stress drawn into his face. I could see the raw panic in his movements as he fumbled through this nightmare, hoping that he might slip away unnoticed. No chance of that here, not when his every move was under the deadly scrutiny of trained watchers.

He crept toward the trunk of a nearby vehicle, his steps clumsy under the strain of gunfire cracking through the air. When he reached the trunk, he grabbed something and bolted south, away from me. He weaved to avoid the blood-soaked figure sprawled on the ground. His attempts at tactical evasion were awkward and amateurish. He faltered as he passed through my line of sight.

I moved my red dot sight ahead of him, steadied my breath, and waited. The moment he came into view, I squeezed the trigger, feeling the dull kick of my rifle as the bullet found its mark. His body jerked violently, his left arm swinging behind his back in a grotesque twist. Blood moved across his shoulders and back as he stumbled, colliding into a low wall. He slowly turned and fell against the wall. He slid down, leaving twin red streaks along the concrete like macabre racing stripes, a brutal symbol of war's harsh finality. Before he reached the ground, I fired twice more. I felt nothing. I was here to do a job, and I was doing it.

During a lull in the fight, I exited the room and entered a large common space. I recognized the Platoon Leader sitting on some stairs and headed toward him. Suddenly, the world rocked, covering me with debris and confusion. My legs felt weak. I blinked, trying to clear my vision. The space around me was a haze of dust, my body numb yet aware of sharp grit embedded in my skin. I forced myself to move, patting down my limbs to check for injuries, my mind struggling to break through the fog of shock. Slowly, shakily, I pushed myself up and made my way back toward the room. The air was thick with the smell of explosives and something burning.

Through the smoke and dust, I caught sight of a Soldier crouched low, his hands pressed tightly to his head as if trying to shut out the turmoil. His face was a mask of fear, his wide eyes searching for something to hold onto, some semblance of control in this hell. His chest rose and fell as his breath came in shallow gasps.

Realizing he was on the edge, teetering between duty and panic, I moved closer, squatting near him, close enough to make sure my words would cut through his shock. My voice stayed calm and measured, as if I were giving instructions on a routine task. "Get to the window and find a target."

He did not respond. Slowly, he shifted onto his stomach, his movements deliberate, almost mechanical. He began to slide across the floor with tactical precision, inching his way to the window, determination gradually replacing the unfiltered terror in his eyes. Once there, he took position. Pressing his rifle against his shoulder, he peered out, scanning his sector with renewed focus.

The smoke continued to drift in thick, choking clouds, and sporadic bursts of gunfire echoed around us. The Soldier at the window became more than a frightened man caught in a firefight. He was a fighter and a survivor, forced to find his resolve amidst the horror and confusion of war. As I took my own position, I felt that same resolve. This is what we had to do to survive this nightmare.

I returned to the room and found AA at the window holding a radio and kneeling in a sea of spent casings. "What the fuck was that?" I asked.

"They detonated the IED," replied AA as casually as reporting the weather.

I positioned myself at a window. Someone screamed out in the street. I scanned for dead insurgents. The one who had slammed into the wall—the one I let live—was still alive. He had managed to exit the fight, possibly crawling over the low wall and into the amusement park.

A few feet away, another insurgent lay sprawled on his back, his lifeless eyes staring up at the sky. A man straddled his waist. He was hammering his fists against the dead man's chest. He screamed in anguish. He gestured frantically, his hands moving from his friend's chest to the heavens above as if cursing the very sky that had witnessed their fate.

I forced myself to watch, even though every part of me wanted to look away. I let the image burn itself into my mind. I had to acknowledge the consequences of my actions and what I had unleashed on this city. Shooting from a distance was easier. You could let the training take over, where each target became a silhouette. But here, in this primal aftermath, I had to face the inhumanity, the agony, and the loss of life.

A group of young men lingered nearby, their attempt at casual movements revealed tension. One of them—a young man in a pale blue long-sleeve shirt—stood out with his crisp clothing and neatly combed hair. I had seen him loitering near the corner before the firefight erupted, barely more than a face in the crowd. But now, something felt off.

I watched him pull a cell phone from his pocket, his hand steady, his head darting around before he lifted it to his ear. Almost as if on cue, a young boy rode up on a bicycle, weaving through the rubble-strewn street. The boy, too, held a phone to his ear, casting glances that seemed too practiced, too deliberate. This was a quiet, orchestrated exchange. They were monitoring us.

I reported the suspicious behavior up the chain of command, my voice low and calm. Orders came back, "Continue to monitor for now."

So, I watched, my eyes locked on every movement, glance, and minute interaction between the two. When the boy finished his call, he lowered the phone from his ear. His face briefly showed something like relief. My attention shifted back to the man in pale blue. He, too, had ended his call and slipped the phone into his pocket, his actions deliberate and calculated.

Over time, scenes like this would become familiar patterns. Their behavior would soon mark them as targets, their fates sealed by practiced suspicion. But today, we only observed them, constrained by caution and protocol. I took notes, mentally cataloging the subtle choreography between them, my senses on high alert. Out here, the smallest hint of intent meant the difference between life and death.

We remained in place until we were ordered to vacate the building and return to base. I was concerned about leaving the comfort of a solid building, walking out in broad daylight, and crossing a four-lane road. I kept thinking, *I am going to get shot, and it will probably hurt like hell as I lay dying.*

We assembled in the courtyard to exit through the gate. As we mentally prepared, I heard a clack against the ground. Someone had dropped their rifle magazine. I thought, *Great, someone is now dropping his gear.*

Hushed words snapped me to attention, "Hey, Sergeant, you dropped your magazine."

I immediately glanced down to find my weapon's magazine glaringly empty. Being left-handed in a right-handed world was a constant challenge, especially during combat operations. My grip on the weapon, with the magazine release button pressed against my magazine pouches, would often lead to an unintentional bump that released the magazine—a frustrating reminder of how the most minor details could complicate life in a world designed for right-handed people.

"What the fuck!" Embarrassed, I picked it up and jammed the thirty-round magazine into the magazine well and pulled on it to ensure a proper lock. It didn't budge. As I thanked the Soldier, someone said, "GO!"

I ran out the gate and into the smoke toward the opposite street corner. *Run faster, fucker.*

Out of breath, I quickly peered around the corner. Standing on the sidewalk was another Soldier, pointing at me to come to him. I did as I was ordered. Nearing his position, I was told to wait and prepare to enter the house. Three other Soldiers arrived. We stacked on the door.

Once more, I heard, "Go." I went.

Inside, I saw people. I scanned for weapons. I heard my own voice informing anyone listening and immediately moved on. "Clear, no weapons, civilians. ROOM! Need another man! Go! CLEAR. CLEAR, CLEAR!" I kept moving. *Slow down,* I told myself. *Slow is smooth, and smooth is fast.*

As if floating, I found myself upstairs, observing the scene through someone else's eyes. I felt like a ghost, drifting in someone else's body, watching hands that were not mine grip the rifle. Another Soldier directed me to the window, his voice distant, almost muffled as if coming from another world. I positioned myself with obedience that felt surreal, a dance I had rehearsed a thousand times, yet somehow happening to someone else.

A shout pierced the fog in my mind—an American voice, sharp and urgent: "MAM (military-aged male) in a black top and bottoms, running left to right." The words rang out like a command, pulling me to the window, compelling me to lean out, to place my cheek against the warm stock of my weapon. It felt natural. I was here to kill this man. Simple, almost automatic, my gaze fixed forward, searching, sweeping until I spotted him.

There he was, darting across my line of sight, his movements quick and clumsy. I scanned his hands with an instinct sharpened by endless training. "He's clean, he's clean!" The words spilled from my mouth before I fully registered them. "He has nothing!" I repeated, almost as if trying to convince myself.

In that strange, frozen moment, I really saw him—a kid, barely fourteen, maybe younger, sprinting away, his small frame swallowed by the vastness of this war-torn city. As I watched him disappear, I felt myself slipping back into the moment, grappling with the blurry line between duty and humanity. Both felt like distant memories. *Focus, Harry.*

When we exited the house, I saw the Bradley Fighting Vehicles (BFVs) were stacked with dead insurgents like dead deer in the back of a pickup truck. One of them was the guy who arrived at the fight with a human shield made of little boys. His multi-colored shirt was hard to forget. Counting the day's KIAs, I felt nothing.

I asked no one in particular, "Why was that guy sitting on top of the dead guy, and what the fuck was he screaming about?"

Someone said, "He was yelling that he was the guy's brother, and he had been walking his sons to school." That didn't make sense. This guy was low behind the boys in a tactical crouch. Why would he insist on taking his sons into a firefight on a street littered with bodies and weapons?

Why? Do you know why? Because it's a lie! That's not why he was there.

The guy I killed died doing what he believed was the right thing. And I killed him doing what I believed was the right thing. This was war.

"The guy I killed died doing what he believed was the right thing. And I killed him doing what I believed was the right thing. This was war."

"All right, get ready to move out." We lined up on the road and hoofed it to base the entire way. I remember getting within 600 meters of the back gate. My M203 felt like it weighed two hundred pounds. At one point, AA verbally pushed me. "Come on, Sergeant. You're all right."

When we got inside, I was exhausted and weak. The heat was killing me—it was 129 degrees Fahrenheit. I dropped my gear, and someone found a cold Gatorade. I reached out for it and wanted to drink the whole thing. Instead, I shared it with the next set of hands. A couple more Gatorades appeared, and the entire process began once more.

After a drink and a short rest, we geared up and moved back to the FMJ building. Everyone was excited. It had been a good day.

Once inside, Staff Sergeant James Gilliland received a report of the day's events. From the deep South, he was kind-natured but also decisive, methodical, and calculated—a true hunter of men.

About this time, I was advised to head to BN Tactical Operations Center because the Colonel had called a meeting to discuss the sniper ambush. I felt a pang of dread. *Was I in trouble?*

Being a sniper comes with its own brand of mental gymnastics. On the one hand, you are putting together a reason to kill someone based on a set of criteria called the Rules of Engagement (ROE). Snipers make judgment calls. They decide life or death for someone, usually before their target takes action. In a normal setting, society would call us assassins, but in war, we are protectors. We must take out the enemy before he takes us out. Because we are making judgment calls, others might second-guess or blame us for a decision made in a matter of seconds. I wasn't sure if that's what I was walking into.

When I arrived, the Company Commanders, S2, S3, FIRES, and Colonel Roggin sat at a large table. He was built like a football player, and his head was completely bald. He had the exterior of a leader, and to his right sat the S3. At the other end of the table, I watched S2 briefly explain the events. On the screen appeared the face of an Iraqi, quickly followed by the picture of the dead Iraqi in a multi-colored shirt.

Oh God, that's why I am here, to explain why I killed a man walking his children. I am fucked.

The S2 officer continued, "Sir, allow me to inform you that our snipers this morning killed the leader of a kidnapping and IED cell. As you can see, we have a perfect match as we overlap the two photos."

Colonel Roggin leaned forward in his chair, strategically placed at the head of the table. He was a human bipedal bull massif who had just been served his favorite dish, dead enemy hajis.

"Goddammit! Yes, results! Gentlemen, I want to hear how these events unfolded before we started shooting. What did we see? Were there any early warning signs?"

I decided to speak up. Without asking permission, I addressed the room clearly and concisely, delivering all the facts. I stayed away from assumptions. I remember mentioning "the totality of the circumstances." The COL looked at me and asked me who I was.

"Sir, I am Sergeant Martinez of the Pennsylvania Army National Guard (PA ARNG). I am attached to your section as a sniper. The first kill was made by one of your snipers." I stopped there and stared at the Colonel.

The COL resumed his position, addressed his staff, and provided some last-minute guidance. "In the future, if any unit needs any additional assets, they are to request them because, by God, they will have them. Am I clear?"

A collective "Yes, Sir!" went up.

The COL stood up and saluted his staff, "Rock of the Marne."

We stood at attention and responded, "Speed and Power."

I wish the story ended there, but days after the event, I kept thinking of those men I killed. They were the enemy. Still, we were in their country, in their neighborhoods, and in their cities. We were foreign soldiers imposing our will on them.

I contacted Delta Company 1SG and requested to speak to a Chaplain. Surprisingly, the 1SG was gracious and made a sincere effort to bring me a Chaplain. A day later, he walked into my hooch. Without revealing anything, he asked me to follow him. We walked to a closed-off area, sectioned off by eight-foot wooden walls constructed of 2x4s and plywood. He escorted me into a large, spacious room with two chairs and oppressive heat.

I sat down, and the 1SG said, "Sit here and wait." He walked out and left me with my thoughts. I wanted to get out of there. I felt I should not display any emotion, and yet, I could see that man falling. My rounds were hitting him as that hand reached out to drag him to cover.

As I got up to leave, another Soldier walked in wearing full battle rattle gear. He had long hair, and he was covered in sweat. His eyes were large and round, and they never stopped moving.

He immediately sat down and introduced himself. "I'm the Chaplain. How can I help you, son?" His eyes were wild, and he could not sit still.

"Hello, sir. I am SGT Martinez, Shadow 4 Team Leader. A couple of days ago, I killed three men and shot two more. More than likely, they died. I am a Catholic and a police officer at home in a quiet town." My eyes began to well with tears, and for the first time, he looked at me. I was falling apart. On the battlefield, my emotions shut down, but they woke up when I got back, and I didn't know what to do about it.

CHAPTER 2
FINDING A PLACE

Patrolman HR Martinez Princetown Township Police Department, 1997

"Hey, you listen to me!" His loud voice was high-pitched and booming. "These men are evil and deserve what they got. I have been here for a year. What you did was right. Don't overthink it. Those five men, as you called them, those motherfuckers deserved it! While you're here, you kill as many of those evil motherfuckers as you can! Are you listening to me? I have watched them kill my men here. I saw them clap in the streets as our soldiers burned or bled out. *You kill them*!"

My composure returned. "Sir. How long have you been here?"

He got out of his chair. "One year too long, and I leave fucking tomorrow." With that, he rose and left the room.

Over the next year, I killed many men. Talking to a Chaplain became my healing process. I never told anyone what I was doing for fear of being relieved of my duties. In doing so, I was able to clear my head and remain mission focused. As soon as I finished, I felt cleared of any sin upon my soul. To this day, I sincerely believe this process also helped me with my transition coming home.

The United States Marine Corps, 1988-1992

Back in 1988, the United States Marine Corps shaved my head, took my civilian clothes, deprived me of sleep, and found interesting ways for me to think, act, and live like an amphibious killing machine. My experience on Parris Island left a lasting mark on my brain and permanently stained my soul.

For the next four years, a transformation occurred that defied the laws of evolution. I was a Marine Infantryman, Special Operations Capable, and assigned to the Fleet Marine Force (FMF), or, as we lovingly called ourselves, "Fighting Mother Fuckers," and I believed it. I was a lean, shooting-trained killer. I trained in hot temperatures with humidity in North Carolina, in the deserts and mountains of California, the jungles of Panama and Honduras, and the trails on the island of Okinawa, not to mention a few bars in Jacksonville, North Carolina.

I discovered I was capable of pushing my body and my mind far beyond what I once believed possible. I climbed mountains with relentless determination, ran endless miles on rugged trails and unforgiving asphalt. I rucked twenty-six miles in under eight hours, defying exhaustion. I crossed deep gorges on single-line ropes, swam through open waters to reach the shore, and emerged ready for any lethal encounter that awaited me. My rifle became an extension of who I was, like a wedding ring is to a groom, symbolizing commitment, precision, and unwavering resolve.

My rifle was with me in every part of the world, loaded with live rounds, except Okinawa where it remained unloaded. I was trained to kill any man or woman considered a threat at the conventional level to a distance of five hundred yards. I was to hunt enemies of the state and kill them. That made me dangerous.

I participated in security operations in Central America and combat operations in the Middle East. I traveled to the South Pacific and trained some more. We ran, rucked, and sailed the seas, conveying a message to the world: we are the US Marines, a shared community that thrives in a world of chaos, discipline, and violence.

In 1990, Saddam Hussein invaded Kuwait, a catastrophic miscalculation on his part. In a matter of days, his once-feared army was reduced to smoldering wreckage along the highways. The infamous "Highway of Death" became a grim symbol of defeat, littered with burning vehicles and looted treasures stolen from Kuwait. What began as a power grab ended as a resounding lesson in the perils of overreaching—courtesy of the United States Marine Corps and the Army.

Before the invasion, Marine Corps leadership identified gaps and voids in the enemy, so we lived in the desert for nine months, training endlessly. We mastered breaching techniques and dead reckoning across a desert landscape with little or no visible key terrain. Even in our sleep, we practiced living and adapting to a chemical MOPP 4-level environment.

MOPP stands for Mission Oriented Protective Posture, and level 4 is the highest level of protection during a chemical, biological, radiological, or nuclear threat. This means all personnel must wear protective gear: overgarment, over boots, protective mask with hood, and gloves. Although this level provides maximum individual protection, it can also reduce performance and cause heat casualties, especially during strenuous activities in temperatures of 130 degrees in the Iraqi desert.

What the world witnessed during those one-hundred hours and what the Iraqis suffered was an instrument of war that had been sharpened to a razor's edge and coupled with a warrior's soul—the United States Marines.

After the First Gulf War, I earned a place with the Marine Surveillance and Target Acquisition Platoon, 1st Battalion, 6th Marine Regiment, 2nd Marine Division (1/6 STA 2nd Mar Div). With this elite group, I returned to my roots in the infantry but at an entirely new level. I pushed myself harder than ever. I grew stronger, ran faster, and honed my ability to multitask under pressure. Through relentless, repetitious training, I learned to trust my gear and sharpened my foundation as an independent thinker.

We practiced jumping out of Zodiacs and swimming to shore to perform a reconnaissance in two-man teams. My equipment weighed me down as the dark, swirling sea tested every ounce of my resolve. Operating as part of a small, tightly knit unit, my confidence grew each week. I imagined the day I would bring our collective tactical expertise to bear on the enemy. We could impose the full weight of war with far more than firepower. We brought sheer will and unwavering discipline. This crucible transformed me, sharpening every edge of my abilities to razor precision.

I got married while in the Marine Corps, and my son, James, was born. He had blond hair and blue eyes. My daughter was born three years later, and she was all me. I eventually separated from The Corps—a difficult decision, but its foundation will stay with me as long as I live.

When I returned to New Jersey, I secured a position as a police officer with the Princeton Township Police Department. Starting in the patrol division, I embraced the challenges of serving my community, honing my skills and instincts on the front lines. Within four years, I earned a chance to train as a sniper on an elite team. Though the sniper role wasn't full-time, it came with a weight of responsibility. Determined to excel, I invested my own time and money into training, and later negotiated with the department to support my efforts.

Over nine weeks of rigorous preparation, I trained in door-breaching tactics to advanced armorer training. Each experience added a new dimension to my skill set, sharpening my capabilities for high-stakes scenarios. By the end of my training, I was ready for the ultimate challenge—a SWAT/ERT sniper callout. The journey from patrol officer to SWAT sniper was a testament to growth, grit, and the relentless pursuit of excellence in the face of responsibility and opportunity.

In 2000, the department's decision to disband SWAT struck me like a blow. Overnight, the pride I once felt as a Princeton patrol officer evaporated. The steady paycheck and excellent benefits were still there, but they no longer filled the void. Something vital was missing—a sense of purpose, of being part of something greater than myself.

Seeking that missing piece, I turned to the New Jersey Army National Guard in 2000. At the time, I didn't know war was just over the horizon. It arrived without warning on September 11, 2001. On a seemingly perfect September morning, four hijacked airliners became weapons of unimaginable destruction, striking the heart of America. Thousands of lives were violently taken in New York, Washington, D.C., and Pennsylvania—a tragedy that would forever be known as 9/11.

Chaos and grief rippled across the nation. The National Guard was called to action to confront the looming specter of further attacks. Despite the urgency and the weight of the moment, my deployment would not come until 2004. In the years leading up to that, the dread of what was to come was a constant reminder of a world forever changed.

Before I went off to serve, another event influenced my direction. In January 2003, on a cold moonlit night with snow covering the ground, I answered the Police dispatcher's radio call: "993, 16. Proceed to the area of Great Road. 911 caller reports…"

A few minutes later, I stood in a Princeton resident's front yard and killed a man as he tried to stab my partner. We tried to control him using voice commands, but he made the fatal mistake of bringing a knife to a gunfight.

While I was serving with the Princeton Township Police Department, the politics of the police department caught up with me. In February 2004, the following conversation took place at Police HQ with my Lieutenant PH, the Operations Officer, about my popularity and the animosity that had grown following my decision to reenlist with NJ Army National Guard.

"Hey, wanker, what's up?" the Operations Officer asked me when I approached him.

I asked him, "Is the Chief mad at me again?"

The Operations Officer said, "Let's go to my office and talk."

Following him, I had this dark feeling that my life was about to change. *Why can't I be boring like everyone else?*

Taking a seat behind his desk, he said, "Close the door behind you."

I felt like a little kid about to share a secret with my older brother. Except this time, I wasn't planning to sneak out of the house and meet some girl with few scruples and less morals.

"Harry, you are the only person I know that can soar like a rocket and crash within seconds of launch."

In a neutral tone, I said, "He's still pissed about the guard thing, huh? You know what? Fuck him. You would think that he would be proud of me, of my service, and trying to better myself. I stay physically fit, and I work the fucking road. I don't hide from work, and I sure as hell don't run from danger like some of the other mopes and—"

He interrupted me, leaned back, and said, "Maybe not, Harry, but the fact remains: Chief hates you. He promised to make your life miserable. He's taking you joining the Guard personally."

My mind went back to a prior incident, when the Chief called me into a room and sitting behind him was his assistant Captain Sunny Perdida with his arm in a cast to discuss my decision to join the Army National Guard. The Chief was sitting in a chair too small for his frame and crossing his arms in front of his chest, full of air and hubris. He said, "Well, Harry, you do what you have to do, and I'll do what I have to do." It was very clear to me he intended to make my life hell.

Returning to our discussion, I remarked, "Let me run something by you. Jersey is going to deploy. I'm not certain when and if we do, but that will flip the old man over the edge."

The officer leaned forward and lowered his voice. Here it came. He asked, "If you deploy, how long will you be gone?" The question hung in the air.

"At least a year."

"A lot can happen in a year," he said, leaning back in his chair. It gave a low creak.

I peered at him, watching for a clue to what he meant. I might as well have been looking at a statue. "What do you know, bro?"

"Nothing. If I did, I would not be at liberty to say anything," he replied, his words carefully chosen.

Proudly, I added, "I will submit the memo later today informing the old fuck that the guard is deploying, further information to follow. What do you think?" I stood up and adjusted my gun belt, an old habit.

As I stood to leave, the Lieutenant said, "I can't tell you to go, but maybe it would be for the best. Don't worry, bro, changes are coming."

In 2004, I was in Sinai, assigned to Alpha Company, 114th Infantry Battalion (1/114th IN BN) 50th Infantry Brigade Combat Team (IBCT) New Jersey Army National Guard (NJ ARNG). As I read the reports about the battle of Falluja, I couldn't help but feel a strong sense of camaraderie with my fellow Soldiers. That's where I needed to be, with them in the thick of it.

I spent nearly eighteen months away from family and my career. By January 2005, NJ ARNG had returned from Sinai. My organization was in the barracks, excited to see our families who were outside, standing in light snow with gray skies hovering above.

The moment we were ordered to move outside, we headed down the hallway, a sea of smiles and anticipation. At the stairs, we ran into a bottleneck, and I called out, "Why are we stopping?"

Someone yelled back, "There's a sign-up sheet. Pennsylvania is deploying to Iraq, and they want volunteers." I could feel the cold coming through the open door. I saw the snow and sleet coming in at an angle. I was ten feet from stepping outside, greeting my children, and returning home to resume my life. In those few seconds, I added my name to the list.

When the time arrived, I would be one of the twenty-five New Jersey Soldiers who joined the Pennsylvania unit, and my life would change forever.

Still on Title 10 federal orders a couple of weeks later, I had another conversation with the same Operations Officer at the Princeton Police HQ. This was in January 2005. The challenges of deployment were coming at me, and I knew I had to be prepared for what was to come. This included my relationship with the Police Department.

The cold, wet, and windy weather matched the starkness of the moment—trees stripped bare of their leaves and grass faded to a lifeless tan. This desolate landscape set the stage for my return to the Police Department. The same office, the same subdued atmosphere, and the same conspiratorial tone shared between brothers in uniform—the place hadn't changed.

I greeted the Lieutenant, "Hey, bro."

Smiling, he stood up to shake my hand. "Hey, wanker, how was it?"

"Just like I told you in the letters." I looked around. "I see the Chief is still here."

"Yeah."

"Is he still angry?"

His expression didn't change. "No. In fact, he liked the CDs you sent. So much he ordered me to put them in your jacket. He couldn't believe you made them."

My chest swelled with pride. "I had a lot of time on my hands. Actually, I was hoping he would retire. What keeps a man here that long?"

The officer replied in an official tone, "That's not for you to figure out. Are you ready to come back to work?"

"Of course. I love being a police officer. But I was a day away from being slapped and out on the street. All those internals and his 'son' were making my life miserable." I was referring to another police officer the Chief protected like he was family. "It will only be a matter of time before I step on my own dick. Hell, I might as well pull it out and say, 'Here you go, step on it.'"

We shared a laugh, but we both knew it was true.

Finally, I said. "The Pennsylvania Guard is short on manpower. They are looking for Infantry to go to Iraq. Word has it that Jersey intends to assist. If not, men will be selected." This was a lie. I said it for my friend's protection. A Police Department can put MTV's *Real World* to shame in terms of drama.

"For how long?"

"Eighteen months, six months training, and twelve months in country. The first month begins as we speak. If I survive, I'll be back in June of 2006." I was looking for encouragement.

Sitting at his desk, this Operations Officer leaned back and said, "I can't tell you to go, but maybe it would be for the best. Don't worry, bro. Changes are coming."

I stood up, automatically smoothing the creases of my black sports jacket. Neither of us grasped the weight of what was unfolding. Six months later, I was at war, killing three men in a single, harrowing day. That quick decision took me through a year of unrelenting violence, its echoes reverberating far beyond what either of us could have imagined.

On March 17, 2005, two days after marrying my girlfriend, I departed for Camp Shelby, Mississippi. Over the next two weeks, we were designated as replacements, moving throughout the camp to fill roles as needed.

The 28th IBCT Pennsylvania Army National Guard (28th IBCT, PA ARNG)

I trained with a new brigade that was short on units and had recently added an armor tank company to its Infantry Battalion, the 109th Infantry Battalion (IN BN). However, the unit did not want to give up its unit designation and decided to keep its original title, A Co 3/103 Armor. This unit suddenly received an infantry mission and had six months to prepare. That might sound overwhelming to the layman, but in all honesty, the company had other hurdles to jump over.

Camp Shelby MS

Since my first deployment with the National Guard, some things had changed for the better, and some problems continued to plague the National Guard during mobilization, especially for Soldiers like me who were attached to National Guard units from other states. That unit does not know the Soldier. He is an outsider. The worst part of almost any organization is the agreements made between Soldiers and commanders (the good ol' boy system), regardless of whether it benefits the unit or not.

The process of assigning Soldiers to units was often random. While sometimes a Soldier's Military Occupational Specialty guided decisions, often they were driven by the immediate needs of the National Guard. Competence. Physical fitness or basic capability seemed secondary concerns. If a Soldier could breathe and pass the Soldier Readiness Process, they were sent forward. This system produced a staggering mix of personnel: the unfit, the elderly, the boastful, the dreamers, and the outright inept. Alongside them stood warriors—some seasoned veterans, others raw but deadly serious about the craft of soldiering.

Every unit had a core group of fighters who shoulder the burden of combat and carry the rest through. Among these individuals, the truth of a timeless maxim shone through:

> Out of every hundred men, ten should not be here,
> eighty are just targets, nine are the real fighters,
> and we are lucky to have them, for they make the battle.
> Ah, but the one, one is a warrior,
> and he will bring the others back.

Those words were a vivid reality in the patchwork of Soldiers I deployed with. Whether seasoned professionals or naturally gifted, a few resolute warriors transformed the motley group into a capable fighting force.

Upon being assigned a unit and receiving an entirely new issue, including uniforms, I immediately pushed to be recognized as a sniper. The roster had me marked as an E5 and a gunner on a high-mobility multipurpose wheeled vehicle known as HMMWV or Humvee. Sitting behind the gun felt like a step down, a hollow cycle with no end in sight.

I could have accepted it as my lot, but deep down, something told me that wouldn't help my growing sense of futility.

To further my frustrations, the traffic control point that we were asked to construct at Camp Shelby did not come close to the reality of Iraq. How we did not lose any more Soldiers than we did that year was an act of God and the flexibility of the Army National Guard. If that had occurred in Iraq, we would have been killed.

I was determined to push for something more. So, I went looking for the Battalion Sergeant Major. I would not recommend my courses of action to the fainthearted.

Convincing myself I had nothing to lose, I walked into the Sergeant Major's office, using the open-door policy. I found him behind his desk, buried in a sea of paperwork, his brow furrowed in relentless focus. The desk told its own story: a fortress of orders, requisitions, and reports that trapped him in an endless routine of logistical warfare. I knew I had to present my case with precision, using my certificates and experience to back up my request. This was a battle of words and will, and I was determined to win.

After a tense exchange marked by stress and calculated persistence, I knew I had only one shot to make my case compelling. I presented a stack of certificates, beginning with my graduation from the USMC Scout Sniper schoolhouse (the law enforcement course) and culminating with my certification as a Police Sniper Instructor. My experience in law enforcement taught me the importance of precision in communication. Addressing superiors in any hierarchical system demands clarity, confidence, and an impact that lingers after the conversation ends.

"Addressing superiors in any hierarchical system demands clarity, confidence, and an impact that lingers after the conversation ends."

With that in mind, I methodically built my argument. I carefully organized each piece of documentation in chronological order, creating a logical narrative of my qualifications and experiences. I wove the facts into a presentation, ensuring every point reinforced my central argument. The final touch was a strong closing statement that left no doubt about my capability and intent. Military leadership values decisiveness and preparation. I leaned into both, knowing that the effectiveness of my approach would determine the impression I left on the chain of command.

The Sergeant Major looked up, took a deep breath, and said," I believe in putting people where they belong. Let me bring this to one of my Joes who has experience in this realm. Together, we will make a decision." His unexpected support was a ray of hope in a challenging situation, providing a sense of relief and encouragement. The gods of war must have smiled on me that day. I had found an unexpected ally.

I sat outside on the only chair available and waited. After an hour or more, I was invited back in. A husky, six-plus-foot giant stood near the Sergeant Major's (SGM) desk with my certificates in his hand. He wore the rank of First Sergeant (1SG). He placed each sheet individually on the SGM's desk while interviewing me.

His words validated my decision to move forward. He informed me that the 109th Battalion had created sniper sections, but the vast majority lacked formal training. He asked if I could help with that. When I confirmed that I could, he exited the office and returned with a tall gentleman, an Officer in an Army uniform displaying a series of additional skill identifiers. On his left shoulder, he wore the coveted Ranger tab.

The officer introduced himself and firmly shook my hand. "My First Sergeant (1SG) informed me that you are a sniper instructor. Let me ask you directly: Can you train my snipers?"

I looked straight back at him and spoke. "Yes, sir."

He asked me to follow him, and we entered his office. He pulled out a binder containing a map. He informed me that we were headed to a hotly contested area called Ar Ramadi. We would be supplementing Marine lines of effort to bring stability to the city as well as destroy Al-Qaeda Iraq (AQI) enemy forces. The coalition forces reported that they endured 60 percent of all enemy attacks in the country. This is as real-world as it can get.

I looked up at him. Trying to project confidence, I said, "I can do it, and I want to lead my own team."

He replied, "I am giving you a sniper section."

When I returned to the company area, someone from the company ran inside the building. My instincts told me it was about me, and I did not get a good feeling. Suddenly, Animal Company First Sergeant (1SG) came out and yelled for a formation.

He looked at me and said, "You stand right there."

Once the company had formed, he lectured them on the chain of command. When he had finished, he dismissed them. He walked by me and said, "Follow me."

He led me to Captain (CPT) Kearny's office, where the Commander sat. I did not care much for theatrics, as I had the sign-off from BN. But I knew to stay quiet, listen, and be prepared to make another brief-but-impactful statement.

Clearly exasperated, CPT Kearny launched into me. He did not yell. He did not have to. His words were sharp and crystal clear.

When I saw the opportunity to speak, I said, "Sir, let me know when I can speak. I have a valid explanation." My voice was firm, and my tone respectful. "You would have done the same thing." Each word was pronounced for effect.

I continued, "I am telling you, I know what I am doing. I am educated in this field. I am telling you, you are going into a terrible place, and you will need people like me. If you were me with my qualifications, you would have done the same thing. That's what we do. We tell you, let me do it."

I stood quietly in front of him, at attention, staring at a space above his head. His First Sergeant (1SG) reflected his anger. He looked like he wanted to tear my head off and was excited to receive the order. I could hear the gallows being built. Despite my mind running away, the silence was deafening in that small room.

I heard his words clear as the sun rose each day: "I will let you do it. But do not ever do that again. You come to me first."

CPT Kearny and his 1SG were a wall of leadership, and I am grateful for that. He granted me the opportunity I sought. It was not handed to me. I fought for it. I wrestled it from the rigidity of the system—and it was mine.

"I fought for it. I wrestled it from the rigidity of the system—and it was mine."

109th's Sniper School

During this time, I met with CPT IN. The S3 Operations Officer and CPT IN became incredibly supportive of my efforts. The second week, I outlined my training program, and he assigned me the training of all the snipers in the battalion, including those from Headquarters and Headquarters Company.

I read books about various snipers from different historical periods and took extensive notes on the lessons learned. I also developed a training curriculum for snipers who had received basic training. I arranged for books to be mailed to me and made phone calls to organizations that supported sniper teams, because snipers often lacked essential gear.

Between March and April, I taught and managed a three-week sniper training school focused on current operational employment. Although I was not allowed to interfere with company training, I was given at least four hours of training each day. The first week primarily consisted of physical training (PT), having the men inventory their equipment, followed by more PT.

I used all available resources. I went to Pennsylvania's Long Range Surveillance Detachment (LRSD) unit and respectfully requested their assistance. These guys were accommodating and instrumental in developing the 109th BN Snipers. One morning, I received a written notification to report to a building in the 110th BN Area of Operations because they were also training snipers. We located the building and sat in an excellent class taught by a well-informed Senior Non-Commissioned Officer (NCO). During the class, I sensed an element of hostility.

The instructor's interaction with us—well, it wasn't there. At break time, I was standing outside, lost in thought, when the Non-Commissioned Officer in Charge called me inside and asked me how we knew about the class. I answered him, and he proceeded to tell me about their selection process. I was happy to hear they had a selection process and asked some questions.

He interrupted me and said in a disgusted tone, "The 109th didn't do a selection process. It assigned Soldiers to the position. My Soldiers competed for their slots. I don't want your Soldiers infecting my Soldiers with their nastiness."

I let him finish, then asked if we could stay for the remainder of class, but I would comply if he wanted us to leave. I was not insulted. In fact, I applauded his brutal honesty. He allowed us to finish the class, and I was grateful for that much.

We continued to train independently and occasionally with the Long Range Surveillance Detachment (LRSD). I could not get the 110's quality instruction out of my head. What a shame. I knew he could teach us some things.

I pressed on. Since day one, I had been asking for time on the range, and after enough pestering, we got it. As a gesture of goodwill, I went to 110's Area of Operations (AO) to locate the NCOIC (Non-Commissioned Officer In Charge) and invited him and his Battalion Sniper section to shoot with us. This small gesture opened some doors for us. Later, we were asked to return to attend his classes. I will always be grateful for his assistance and Sergeant Wertz of the LRSD of Pennsylvania and his team's valuable input.

In addition to their training, I taught the 109 everything I had learned about marksmanship, field craft, tactics, techniques, procedures, and rules of engagement from the Marine Corps and as a police officer. We also did a lot of PT. I held nothing back in training. I shared the mistakes I often made while shooting and the things that worked for me. The class started with twenty-seven guys and ended with twenty-one.

I was now part of Alpha Company, 3rd Battalion, 103rd Armor Regiment, 109th Infantry Battalion, 28th Infantry Brigade Combat Team, and the section leader of a seven-man sniper section.

In May 2005, we went to the National Training Center. One of my NCO's informed me that "he felt he was out of his league." I listened and thanked him for trying and walked away. Two more left soon after that.

We did eleven days of training at Fort Irwin. That ate up the entire month between transportation, signing out the vehicles, and moving out to the training site. We did not train at night using NODs (Night Observation Devices). We did not do any PT. We did not practice anything related to Iraq Operations. Traffic Control Points were conducted at night with lights on. The training was inferior and impractical.

Kuwait/Iraq

We finally began moving to Iraq in June of 2005 and arrived in Kuwait City. The last time I saw this city, it smelled of burnt oil. It was partially destroyed and occupied by an invasion force. The Iraqi Army was fleeing north to escape relentless U.S. Marine forces.

At Camp Buehring Kuwait, we continued to process and attended some training cycles in a hurry-up-and-wait period. After what felt like weeks, we moved to the airfield for departure into Iraqi airspace. Our destination was Marine country: Al Taqaddum Air Base (TQ), an

Arabic term meaning progress. Today, the base is manned by Iraqi security forces. After a few days at TQ, we boarded open-back vehicles in a convoy to Camp Corregidor Ar Ramadi, a city in the heart of the Sunni Triangle.

We arrived at Camp Corregidor in the middle of the night.

CHAPTER 3

CAMP CORREGIDOR

USMC 1st ANGLICO Wild Eagle 5-7 CPT Jason Angell and his team, Hit, Iraq 2005

3rd Infantry Division, Rock of the Marne

Between 2005 and 2006, Camp Corregidor was one of the most austere and dangerous outposts for Coalition forces on the outskirts of

Ramadi. The camp was strategically placed to protect Ramadi's eastern neighborhoods and surrounding areas.

Camp Corregidor was originally an agricultural college surrounded by high walls that were now reinforced with Hesco barriers, which are basically sandbags on steroids. Along the walls were towers, some already in place and others constructed from building materials. Razor wire snaked around its perimeter. While I was there, these positions were enhanced, such as bullet-proof glass retrieved from partially destroyed military vehicles. These positions were often attacked by enemy snipers and rockets. Its concrete barriers were a constant reminder of the threat hovering outside.

Living conditions were harsh with limited access to running water and air conditioning in the blistering heat. Soldiers slept in cramped quarters on makeshift bunks with the ever-present hum of generators. These rooms once held classes filled with students.

Camp Corregidor was a target. Insurgents regularly attacked with mortars, rockets, and small-arms fire. Sniper activity in the surrounding urban terrain added to the constant tension. Close to insurgent strongholds, the camp was a critical launching point for combat operations—counter-IED missions, sniper overwatches, and direct assaults on insurgent positions.

The Soldiers stationed there lived under the physical dangers of combat as well as intense psychological strain. We had to pause, check the environment, observe how other people were moving, then sprint from building to building to avoid the constant threat of gunfire waiting to mow us down.

One time, I was on my way back from the mess hall when a mortar round landed. I heard a second one and knew I had twenty-six seconds to find a place to hide. The only place I could find was a cement barrier that was so narrow I could only fit my head and upper body into it. My butt was sticking out. The mortar landed and didn't explode. Standing up, I saw its fins coming out of the ground.

In these stressful situations, I always went back to my boyhood hero, Chuck Yeager, who stayed calm no matter what. My hands were shaking so badly, I could hardly get out my radio. In my mind, I kept repeating, *You're Chuck Yeager. You're Chuck Yeager.* I reported the unexploded ordinance like announcing the weather report. My hands were shaking so much that I couldn't clip the radio back to my gear, so I dropped it into my pocket.

Despite relentless attacks, troops maintained their resolve. Camp Corregidor was a symbol of Coalition defiance. It was a foothold into Al-Qaeda in Iraq's operations. For those who lived and fought there, the camp was a place of camaraderie forged in hardship, a stark reminder of the sacrifices and challenges of war.

Late one night, we were briefed by the commander of Death Dealer Company 2nd Battalion, 69th Regiment. Scanning the room, I noticed that I knew no one who would risk his life for me. The regular Army Captain (CPT) was middle-aged, pale, and average height. He said some words, and when he was finished, the Commander (CDR) spoke of his convictions, "bringing the fight to the enemy, and I believe in the cause…"

Keeping up an appearance of respect, I was busy selecting a temporary sleeping spot in the classroom. *Ah-hah, a good spot is directly beneath the fan. I hope it works.*

Finally, the CDR said, "Again, welcome to Corregidor, and we look forward to working together—One Team!"

Next to him stood a large African American Soldier. Everything about him screamed business. He provided us with a small verbal picture of his past and introduced himself as the 1SG of Death Dealer (2/69 3rd ID MECH). He ended with, "I love the Army. That's why I never got married…," and he meant it.

Searching for a suitable cot, I spotted one. It wasn't perfect, but it was the cleanest option available—or close enough. Ignoring the faint stains and hoping they were only sweat, I claimed it. Once the 1SG wrapped up his spirited hooah speech, I tossed my gear onto the cot with a satisfying thud and stepped outside into the cool embrace of the night air. I craved alone time.

Without warning or explanation, I had just learned that my entire sniper section had quit. The betrayal cut deep. My three NCOs had abandoned the mission. One claimed he was out of his depth. Another couldn't bear the thought of spending the tour apart from his girlfriend, and the last, astonishingly, declared he wanted to spend the rest of his deployment sitting in a tower on Force Protection duty. Adding insult to injury, he was the nephew of the Animal Company 1SG.

Their departure rippled through the ranks. Their subordinates, the Specialist E-4s who had once looked to them for guidance, followed suit, leaving me with nothing but the hollow silence of an empty section. The weight of their choices hit me hard—a mixture of shock, betrayal, and disbelief. I had trusted them, depended on them, and they had walked away.

The entire battalion-sized Forward Operating Base was under blackout conditions, a precaution against the ever-watchful eyes of enemy snipers. The darkness was thick and oppressive, yet strangely serene. I took my time navigating through it, letting the quiet settle over me like a blanket. Finding a crude bench nearby, I eased myself down, the weight of the day momentarily lifting.

I sat there, still and patient, waiting for my eyes to adjust to the shadowy void, smelling earth. An acrid charcoal scent mixed with a whiff of the latrines down the way, and always the constant drone of the generators. Soon, faint shapes began to take form. A dark shadow emerged in the distance, moving cautiously and deliberately. I chuckled softly to myself, watching the figure's wary approach, every step a dance of hesitation. In that moment, amidst the tension and the quiet, I felt a strange comfort in the familiar rhythm of war and our shared vigilance.

Keeping my voice low, I called to the Soldier, "Blink your eyes a couple of times, slowly, and you'll be able to see better."

I turned my focus back to my surroundings, brushing off the Soldier's presence. I just had the rug pulled out from under me. I needed a new plan.

Thrown in with the 2nd Platoon under the command of a young LT Medina and a tall, squirrelly-looking Platoon Sergeant (PSG). I was a Sergeant E-5 with Staff Sergeant E-6 responsibilities, about to be reduced to Specialist E-4 duties. A wave of self-doubt washed over me. *Be cool, Harry. No one has said a thing to you. Pack your gear and think it over in the morning.*

After collecting my footlocker, two duffel bags, and a rucksack, I returned to my sweat-stained cot. I spread out my poncho liner and my sleeping bag, put together a hasty pillow, and thought, *Fuck, what do I do now? I hate towers…!* as I drifted off to sleep.

The following day, I made my way around Camp Corregidor to check out my surroundings. My assessment matched my opinion from last night. *What a shithole.*

I headed to breakfast inside a hot, fly-infested building. Breakfast resembled two elements of solid human waste and last night's regurgitated dinner. I studied my plate for about five seconds and dumped the foul-smelling stuff into the full garbage can located outside the building.

"SGT Martinez!"

"Good morning, SGT," I responded. He stood as tall as some girls I knew, and his demeanor matched the same gender. His most defining male trait was his mustache.

"Go see the Platoon Sergeant as soon as you get done eating."

"Roger, I…"

He was already walking away. I would have to find him on my own. As I walked into Animal Co's Area of Operations, I came face-to-face with former team members. They looked away from me as if some unseen force had diverted their attention. *Shit.*

I entered a building that had once taught students agricultural science. A quick view of the desert indicated the learning curve. Outside the entrance hung a fresh aroma of sewage with the buzzing of its companion, flies. Four wooden porta-johns stood conveniently located outside the main door.

Inside the building, the walls were textured plaster, scarred by weapon barrels and furniture. To my right was the entrance to Morale Welfare and Recreation, known affectionately as MWR. To my immediate left were the dorm rooms of Animal Co.

Some Soldiers lounged in the hallway. Others were moving with purpose. I tried to act indifferent and said hello to anyone I recognized. I stepped into a room that resembled a TOC (Tactical Operations Center).

"Has anyone seen the 2nd Platoon's PSG?" I called to no one in particular. My voice hung in the air for a moment before the readiness NCO turned to face me. His expression carried a smugness that instantly set my teeth on edge. When he finally spoke, his voice was low, almost a mutter, dripping with condescension as though the effort of addressing me was beneath him.

I sized him up. To me, he represented everything wrong with the Army's standards—or lack thereof. His meticulously regulated haircut framed beady blue eyes that darted around with unwarranted self-importance. Above his upper lip, a poorly trimmed mustache was cut so short it almost seemed a mockery.

The most prominent feature of this man wasn't his haircut or his mustache. His waistline bulged beneath his uniform, a glaring contradiction to his air of superiority. His ego matched his girth. A wave of disgust churned in my stomach, not just for his appearance but for the attitude he carried like a badge of honor. Arrogance made my skin crawl. It had no place in a war zone.

This was the norm in the National Guard during this period. The Global War on Terrorism would later change this, as the Army National Guard moved from a Force in Reserve to a Force in Operation.

Twitching his moustache, the NCO said, "He's probably in his room at Delta Barracks."

"Roger that, thanks." I walked back out and headed to Delta barracks where I also slept. I found his room and knocked three times.

"Yeah." A frail voice came through the door.

"It's SGT Martinez."

"Come in."

I wanted to report formally but didn't think he deserved that much. Instead, I said, "You were looking for me?"

He lay in his bed, looking extremely comfortable in his two-man room with AC, reading a novel. "Yeah, you are on towers tonight. I haven't assigned you to any squad, so stay flexible. The schedule is on the erasure board."

"Cool Man, thanks, SGT," *for wasting a year of my life in towers, you fucking clown.*

I wanted to sleep that night, but instead, I shared the towers with another Guard Non-Commissioned Officer (NCO). We made the best of it for four hours. *How would I get any combat operations as a sniper in a tower?*

The next morning, LT Medina approached my rack and informed me that since they had no space for me in an Armor Tank/Infantry company, I should look for 503rd 2nd Infantry Division snipers operating under the name Ghost. LT Medina appeared sincere in his message, so I thanked him and searched for CPT Kearny, my company commander.

After locating Animal Company, he informed me of a sniper section on Corregidor. My task was to find them and try to get a "lesson learned" brief from their sniper section leader. I thanked the CPT and smiled inside. Not all was lost. I had to separate myself from Delta Co. One element was missing, though. I needed a partner. I decided to bring SPC JY into the fold.

SPC JY was one of Pennsylvania's National Guardsmen, a quiet twenty-year-old college guy majoring in biology, now thrust into the madness of war. With his brown hair, sharp features, and striking presence, he carried an air of quiet resilience, though his reserved nature often tested my patience. "Speak up, JY," I'd joke. "This old man's deaf!" I was thirty-eight at the time and felt ancient next to these young guys.

I found SPC JY in his room. It was a classroom converted into military billets for 3rd Infantry Division Armor of the 2nd Battalion/69th Regiment. He shared his room with eight other Soldiers. I was reluctant to speak to him. I did not know him or how he felt about me leading a sniper team. I knew from others that SPC JY and his peers had been warned to stay away from me. I figured the best way to approach this situation was to feed his need to be a sniper.

I did most of the talking during the conversation, and SPC JY listened. If anything could be said of him, he was a man of few words.

I said, "JY, I want one thing from you: your loyalty. Give me that, and I will lead you into a shooting war. Our job is to locate the enemy and use long-range precision fire to kill them. The Iraqis understand that positive or negative action does not matter. If the Iraqi is sitting on the fence, and our actions cause him to spill to the dark side, so be it. He just made our job easier." I paused and stared into his eyes. He was taller than me, but that didn't matter. "Do you want to do this? Do I have your loyalty?"

SPC JY's face showed no emotion. He stared at me with the same degree of intensity and replied, "Yes."

We were on our way.

I found Ghost Team on the southern end of the FOB (Forward Operating Base). After entering the living quarters, I found the team asleep. Initially, I was surprised, but later I would do the same. The American Army operates at night, hence the sleep cycle.

Eventually, I made enough noise, and a Soldier propped himself on his elbow. I introduced myself and explained my purpose. He rubbed his eyes, trying to wake up.

"Lessons learned, huh? Don't go out in two-man teams. They will kill you." With that, he dropped his head on his pillow, instantly asleep. I looked at SPC JY and glanced at the door. In the hall outside, I kept thinking, *I'm fucked.* Aloud, I said, "Give me some time. Head back to your room. Those Soldiers are more than likely burnt out. Later."

I returned to the CPT's sleep area and reluctantly reported my findings. Despite the obvious disappointment, the CPT added, "I heard there is another team here from 2/69, led by Staff Sergeant (SSG) Gilliand, operating under the name Shadow. Find them and see what they tell you.

I felt inspired and did not delay. Crossing Camp Corregidor, I went straight to the building referred to as Full Metal Jacket (FMJ), a two-story building partially destroyed and at the same time under construction. Like all buildings in Ramadi, it was made of cement brick. At one time, it might have housed the administration office.

As I approached the building, I saw a hole on the south side, large enough to allow people to walk in unencumbered with their equipment. Leading into the hole was a set of wooden stairs. I knew a sniper section was living in the building, and I was to locate SSG Gilliland. I kept thinking about what to say while remaining "relaxed, confident, and humble."

As I stepped into the hole, I smelled paint. Two Soldiers wore Desert Camouflage Uniform bottoms and brown T-shirts. They were painting the M24 Sniper Weapon System. I stood there and watched briefly. The shorter man was stocky with a sober, controlled demeanor. The other one wore glasses. When the stocky man made a motion for me to pass, I said, "Sorry to bother you, but I'm looking for an SSG Gilliland.

The taller of the two spoke up. "He's not here right now. Can I help with something?"

"My name is SGT Harry Martinez. I'm assigned to the National Guard unit that arrived the other night. I am in charge of a sniper section, and I am looking for SSG Gilliland to coordinate training and, if needed, assist on a mission.

The shorter man spoke up. "How large is your section?"

Oh shit. Might as well spill it. "This unit was a tank unit, now reclassified as an Infantry company. My sniper section initially consisted of seven Soldiers, but last night, all my NCOs quit. I have three E-4s, so there are four of us.

The Soldier wearing glasses glanced at the stocky one.

Desperately, I added, "Gents, I'm looking to work with you. I am an infantryman, trained in the Marine Corps. My last unit in the Corps was a surveillance target acquisition platoon. I am also trained by the FBI as a Police Sniper. I have been training for this position since 1991.

After a slow, agonizing moment, the bigger one extended his hand. "I'm SGT SH, team leader from Shadow 1. This is SPC Collette. Give me a moment." He pulled out his radio and said, "Shadow 6."

Shadow 6 responded and made some inquiries.

When they finished their conversation, I informed SGT SH. "All I have for weapons is my M16A4 with an Aimpoint and my partner, SPC JY."

SGT SH asked, "Do you want to go on a mission with us?"

Surprised, I heard myself reply, "Yes."

"Great. Be here with your Soldier at 2400. It's a twenty-four-hour mission."

This informal manner of Shadow Operations would be its signature on all operations involving Shadow Team… and I loved it.

The First Mission

That night, I methodically packed my assault pack: bottles of water, a few snacks to keep my energy up, and extra magazines, each meticulously loaded with 29 rounds. My M16A4 rested by my side, equipped with an Aimpoint sight for precision. Its weight felt familiar and reassuring. Two hand grenades tucked securely into my gear were a silent reminder of the stakes.

As I stepped into the dimly lit hooch belonging to Shadow Team, the room carried the faint smell of cleaning solvents and sweat—pre-mission rituals in full swing. My gaze settled on Staff Sergeant James Gilliland, better known as Shadow 6. He was strikingly composed, his sharp greenish-blue eyes alive beneath his beanie that barely concealed his short brown hair. He stood a few inches taller than me, his presence commanding yet approachable.

His steady, assured posture told the depth of his experience. When he spoke, his smooth Southern drawl carried calm authority, cutting through the pressure with a disarming warmth that felt out of place amidst the cold precision of our preparations. Shadow 6 was the kind of Soldier who made you believe in the mission, even when doubt lingered in the back of your mind. He was a warrior.

I introduced myself and SPC JY, my comrade for the night. SGT SH barked out a set of instructions with the tone of a man who had done this hundreds of times, knowing each time could be the last. We nodded, double-checking our gear under the red glow of low-light lamps, then moved out into the night.

The North Gate was our exit point, a looming metal barrier that separated the relative safety of the base from the unknown in Ramadi's hostile streets. Joining an infantry squad, SPC JY and I went out with SGT SH and Shadow 1. We might be new to Ramadi, but we were already well-oiled machines ready to bring destruction to the enemy. That morning, I killed the man who surrounded himself with boys, and I let one get away. We finished the day with 3 verified KIAs.

Shadow 4

After the success of our first mission, we were pumped to go out again. SSG Gilliland asked me how many more Soldiers were trained as snipers. I informed him that besides SPC JY, there were two more, SPC Bennett and SPC RT.

"Good, well, bring them over. They will be your team, and you will be called Shadow 4."

I nodded and thanked him for the opportunity. I was being a little reckless. I had yet to speak to Bennett and SPC RT, and I had no idea if they would agree to work with me.

Next, I hit the shower. Grit had turned the hems of my shirt sleeves into sandpaper. I still had grit in my teeth from the IED that rocked our outpost. The cold water felt good, but my face burned with dozens of nicks from the blast. The second I turned off the water, I started to sweat again and had trouble getting my clothes on. Everything felt sticky.

As I sat on my bed, lost in the rhythm of my own thoughts, Bennett's voice carried over the low hum of the large room. At first, I barely registered the words, but his frustrated tone caught my attention. He felt let down by his NCOs since the Pennsylvania Army National Guard (PA ARNG) sniper section would not be operating as a tank platoon.

I stood up, brushing aside the poncho I had draped to create a makeshift barrier between myself and the others. Isolation was my refuge, a way to shield myself from the weight of camaraderie in a place where loss was a constant threat. For a moment, I stood there, hands on hips, listening but saying nothing. A fire stirred within me, part frustration, part anger, but also recognition of the same doubt that had been eating at me in quieter moments. Bennett's voice was a mirror of the mounting discontent that hovered over all of us.

I strode to him in a police-like manner and spoke in my police command voice. "Bennett, let's go outside and talk." Once we were in the hallway, I channeled him against the wall and tore into him. "I have a question. Give it thought before you answer. I came here to fight. I don't care about winning hearts and minds. That being said, I'm not going to kill random people either. We will act in accordance with the Rules of Engagement (ROE). While here and between missions, we will train hard. I am in charge. I expect you to be loyal to me. I need to trust you on every level. If you want to run and gun, I will allow you to join my team. If you decide not to, then stay the fuck out of my room because I do not want to hear your voice."

Bennett stared straight ahead and replied in a stern voice, "SGT Martinez, I want to join your team. I spoke to SPC JY, and he told me what happened when you went out. I want to be involved in that." A smile slowly crept along his face.

I remained stone cold. "Bennett, I want to hear an apology and say you will be loyal. Otherwise, go fuck yourself. Oh, and if you undermine me. I will fucking kill you myself."

Awkward silence, then, "I apologize, SGT Martinez, and I will be loyal to you and the team."

I eased back, relieved but not showing it. "Good enough. Return to your room and ask SPC RT if he wants back. I have the same expectation of his conduct. Explain it all to him. I will be by later to hear his answer. Are we clear?"

"Yes." Bennett could not contain himself. He had the smile of a kid opening his Christmas gift.

I walked away and sat on my bunk. *Two down, one to go.*

To kill time, I walked over to FMJ to locate SSG Gilliand, but he was at the HQ building. I met him as he was exiting, and said, "Hey, SSG Gilliand. My team is ready to roll."

He was in a good mood. "I just spoke to my Battalion Executive Officer," he said. "You and your team will now be assigned to HQ 2/69 Sniper Section. Furthermore, I understand you are actually a section leader. Can you bring in more guys?"

I said, "No, I'm fine with these three. In my professional opinion, the others are better off where they are. I am fine being a team leader with these Soldiers."

Without hesitation, SSG Gilliland stated that I would operate a four-man sniper team. The new Army sniper field manual would later adopt and practice this formation, and it would come to be called a heavy sniper team.

That afternoon, I approached SPC RT in his room with Bennett present. After his apology and his promise to be loyal, Shadow 4 was ready to begin operations. We would accompany three different teams, and I had no issue with that. After all, they had to get to know us and vice versa. Following the success of our first two patrols that resulted in enemy KIA, the enemy tempo slowed down, showing a new player was in town. 2/69 was stacking dead insurgents.

Shadow 4 got together in the morning to do physical training (PT), followed by range time on a hundred-meter range, enabling us to focus on fundamentals. Afterward, we ate breakfast. For the rest of the day, we devoted ourselves to taking care of our equipment and learning about Shadow's TTPs (Tactics, Techniques, and Procedures). I continued to read current after-action reports and make notes. I asked Shadow team leaders many questions, and we all reviewed the maps.

One area was circled and labeled The Badlands. This area interested me, but I kept my thoughts to myself. SSG Gilliland allowed his team leaders a great amount of flexibility as long as we kept him informed. At times, he would intervene but only for professional development. I liked the concept and found his style rewarding.

Shadow 4 had two functioning M24 sniper rifles. Divided into two teams within Shadow 4 would be Bennett and SPC RT, then SPC JY and me. This was based on personality and compatibility, in addition to living arrangements. JY and I lived in the same building, while Bennett and RT lived in another building with rooms across from each other. This allowed the two-man teams to build a bond and grow trust with each other. Unknowingly, this space between us would allow us breathing space and help us manage any drama between us. Looking back, we were extremely fortunate on so many levels.

Most of the time, we went out on missions separately. SPC JY and I would go out with one unit, and Bennet and SPC RT would go with another. Sometimes, only one man would go. We had to stay flexible because every day was different. Because of this, we got to know a wide range of the leadership and also the men in the various units.

Shortly after arriving, I was in the chow hall when a dark-haired guy with glasses took a seat across from me. He had a quiet, friendly demeanor. He introduced himself as Sam and told me he was an Arabic interpreter. His parents were Iraqi immigrants to the U.S. Like me, he was a second-generation American. We often had long discussions about Iraqi culture. Sam helped me understand the local people and their ways.

As the weeks continued into August, I found friends among the Shadow team. These guys were studs, but no one was perfect. Without a doubt, that included me. We were all independent thinkers, but when it came to time to leave the wire, we became a single-focused fighting machine.

Besides training, I familiarized myself with the city by using the map and conducting recon patrols at night. Using NODs might appear to be an easy task, but in reality, getting used to them took some time, overcoming the desire to flip them up, as well as adjusting to their annoying weight compressing the back of your neck while running.

Our missions were dictated by two critical factors: the need for water and the ability to remain concealed within the city. Each mission typically lasted around twenty-four hours. For example, we might leave the wire at 0330 on a Tuesday, though it still felt like Monday night. The goal was to be in position before the morning prayer, using the cover of darkness to our advantage. We'd remain in place, working throughout the day, observing and reporting activity into the early hours of Wednesday, moving again after curfew. This timing took advantage of the enemy's fatigue and complacency during those hours, at least that was the idea.

Following a shooting incident, we would be extracted or conjure up a plan to exfil back to Camp Corregidor. Shooting was only one part of the sniper training. We had to master fieldcraft in locating a suitable place to watch from concealment and reach that place without detection. Afterward, we had to get out alive. We would usually patrol our way back to base, hopefully unseen. On other occasions, we would have to be extracted from an enemy actively trying to kill us. As the leader of Shadow 4, I had to plan and execute all of it.

We avoided following a pattern to keep the enemy from discovering trends. Watching the Shadow team work at night, moving among a hostile populace was an art. Approaching corners was similar to looking both ways before crossing the street. Urban warfare was a 360-degree-war where we also had to look above and below. This heightened sense of vigilance was physically and mentally exhausting and, at times, comical.

As the months went by, I began rolling my ankle and falling, often as we neared the exit point during operations. At first, we laughed it off, but over time, it became a significant concern—not just for me, but for the entire team. To address this issue, I incorporated intense exercise into my routine, focusing on squatting and completing 10x10 sets of pull-ups to build strength and stability. By the end of my tour in Iraq, my efforts paid off, reducing the frequency of rolled ankles from a daily occurrence to just once in the final six months.

I'd get emails from my family, but the idea of home became like another life, almost forgotten. I accepted this rugged, dangerous routine as my new normal. Although I did not want to die, I made the decision to put death out of my mind as a way of maintaining stability and focus. I had no idea how difficult the shift back to my home and family would be after this complete immersion into the inhuman hell of Ramadi.

Despite my intention to forget, thoughts of my family and my growing-up years still came to me while I was doing PT or trying to get some shut-eye.

CHAPTER 4

HOW I GOT HERE

My parents, Carlos E. Martinez Aguirre and
Angelica Rosa Martinez on their wedding day, Bogota, Colombia 1962

December 1965

In 1964, my parents, Carlos E. Martinez Aguirre and Angelica Rosa Martinez, came to the U.S. from Colombia. They had virtually no education but plenty of determination and courage. My father never spoke of his past, but history tells us of internal conflict in Colombia from 1948 to 1958 that claimed an estimated 200,000 to 300,000 lives and displaced millions, leaving Colombia in a state of struggle that lasted for decades.

This young couple made the decision to abandon all they knew and depart with nothing but optimism in their back pocket and with my older brother, Carlos in their arms. My father navigated the complexities of obtaining a visa and secured employment in a distant land. His resolute spirit brought our family to New Brunswick, New Jersey, and the promise of a fresh start.

In 1965, I was born in New Brunswick while my parents were grappling with a foreign culture and navigating the upheaval in the U.S. during the 1960s, including civil unrest, the Vietnam War, and cultural change. Within four more years, my mother would bring my sister and younger brother into our world.

My father stood five feet seven, a commanding presence with his loose, curly hair slicked back and a neat mustache. He had a muscular build bronzed by the sun from his daily grind under open skies. He worked tirelessly to grab his piece of the American dream while assimilating his family into a new culture and mastering a foreign language. Inside our home, Spanish flowed freely, a comforting link to our roots, but the moment we stepped outside, English dominated every word, a practice that mirrored his relentless drive to adapt and succeed.

He was a man of quiet pride, a provider beyond compare with natural leadership that inspired respect. Sharp and resourceful, he quickly learned to speak, read, and write English, a skill that opened doors to opportunity. In no time, he launched a fencing company, a gas station, and a taxi business. While nurturing was not his strength, his role as a provider was unmatched. We didn't have much, but thanks to his efforts, we never went without.

My mother complemented him with her unwavering discipline and vision for our future. She instilled in us a clear path to adulthood with expectations as rigid as her rules. Straying even slightly from her vision meant facing consequences that might seem severe by today's standards. By the age of twenty-six, with my youngest brother in her arms, she was a force to be reckoned with—cursing like a poet and running her household with an iron fist wrapped in the deepest love. Together, my parents created a home with a legacy of resilience, sacrifice, and determination.

Around fourth grade, we moved to South Brunswick into a brand-new housing development. Every day, new families moved in. The majority were Italian and Irish, but we also had Jewish, Hispanic, and African American families. Down the road at a T-intersection stood St. Augustine of Canterbury Catholic Church.

I started out in the public school system, but my mother convinced my father that her children needed to be close to God and attend the school at St. Augustine. After several heated debates, my father finally relented. No more rugged tough-skin jeans from Sears. Now, I dressed in sleek corduroy Levi's with a pale blue shirt and a neatly knotted blue tie. Ironically, despite my struggles in school, those moments of preparation for middle school and high school became some of the happiest days of my youth. They marked a quiet transformation, filled with hope and a sense of possibility that overshadowed the weight of my academic shortcomings.

After St Augustine's, I attended St. Joesph's Brothers of the Sacred Heart, an all-boy school (Class of 1984). St. Joseph High School maintained rigorous academic standards with a robust curriculum to prepare students for higher education. Their students came from all over the state to compete with life, themselves, and each other.

Between my junior and senior years, I made a life-changing decision and chose Thanksgiving to tell my family. Using her poetic ability with colorful language, my mother advised me in Spanish, "You better not fucking ruin Thanksgiving for the family."

Against her advice, I decided that was a perfect time to inform my father that I had no desire to attend college, and my goal was to join the military and later become a police officer. She was right. I ruined that holiday for all of us. My father was a quiet, angry man. He never yelled or laid a hand on me. He spoke, and we listened. Anything less was disrespect. My mother was the one who employed child training in a Colombian manner.

Almost every event in our house had a dress code. We would dress in suits and ties on special dinners such as Thanksgiving and Christmas. My father always bought us custom-fitted three-piece suits in East Brunswick from an Italian tailor along the Route 18 corridor.

He always sat at the table with his sons on his right and my mother at the opposite end of the massive dark wood table with high back chairs and plush red carpet. She would supervise my younger brother and sister. To say the least, these events were formal, and they established a protocol that we followed into our adulthood. I still dress up for events. Respecting traditions allowed me to readily adopt other traditions that would shape me into a warrior.

When I told my family of my decision, I broke family protocol. My mother got up from the table. She left the house and took my siblings with her. Considering my parents' experience in Colombia with the police and the military, I was taking a step in the wrong direction. They believed I would do better as a lawyer or something equal to that position. Over time, their views would change, and they saw the value in serving the country.

Nothing deterred me from my decision. I joined the New Jersey Army National Guard. During my junior year summer, weighing 120 lbs., I boarded a plane, landed in South Carolina, and rode a bus to Fort Jackson for Army basic training. I returned home in late August and resumed my senior year. That year, I attended weekend drills.

My mother asked me what it was like, and I gave an answer that shocked her. "It's like Boy Scouts, except we have more formations. They drink, some get high, and it becomes a big barbeque weekend." I added, "Mom, how we win wars with this group is beyond me."

I graduated at the bottom of my class and felt I had embarrassed my parents. After the ceremony, I left with my girlfriend to go to the Jersey Shore for a couple of days. I did not want to deal with my parents' wrath. A couple of weeks later, I boarded a plane bound for Fort Gordon, Georgia, and attended school to study for the position of combat teletype operator. I graduated with honors. When I returned home, I shared the letter and certificate with my parents.

While at the Signal Corps school, I learned that life is filled with contradictions that challenge the very core of our beliefs. In one of his quieter moments, my father warned me, "Friends appear as individuals," he had said, "but they come and go like the seasons, each with their own agenda, right or wrong. They will judge you, as you will judge them. Among these, you may find two or three who become closer than brothers but never more than that number."

I learned this hard lesson at Fort Gordon when I became friends with Edgar who was a college man. We often shared laughs, comical stories, and dreams about the future. I saw him as a close friend.

When I learned I would be an honor graduate, I headed to Edgar's room to celebrate with him. Smiling and excited, I knocked on his door. He frowned when I blurted out my good news. Instead of letting me in, he insulted my heritage and shut the door in my face.

Walking away, I knew my dad was right. Life is neither fair nor gentle, and not all friendships are built to last.

> "Life is neither fair nor gentle, and
> not all friendships are built to last."

I returned home to a job and another girlfriend and became a part-time student, part-time father, part-time brother, and part-time son. Nothing motivated me. I was all over the place, filled with distractions, poor decisions, and frustrations. I floated through life for five years, trying to make sense of it and put my life into some kind of order.

I was still serving with the New Jersey Army National Guard when I was selected for Sergeant (E-5) after successfully passing the promotion board. Assigned to Charlie Company, 250th Signal Battalion, 50th Armor Division, I held the position of section leader. This was something I could do relatively well, but it was still not enough. I craved a challenge, something that would push me beyond the ordinary. That drive led me to Fort McClellan, where I attended the Military Police School, a Regular Army course. I completed the course, but instead of moving forward, I returned to find my orders for promotion rescinded. I was back to being an E-4. By 1987, restless boredom took hold of me.

One rainy morning, that sense of stagnation came to a head. While driving to work, I was pulled over by a police officer and handed four summonses—each one a sharp memento of my laziness and carelessness. I sat in my car, frustration boiling. Light rain streaked down the windows. Staring into space, my eyes eventually focused on an enormous billboard. Against a black background, a Marine stood proud, his white cap and sword slicing his image into light and shadow. He meant more than a man in uniform. He was challenge personified. Below him, bold letters read: "The Few. The Proud. The Marines."

My loyalty was with the Army, so I visited an active duty (AD) recruiter. They informed me that the Regular Army would accept me, but my rank would be reduced to E2. They would only recognize my basic training and the AD schools I had attended. In 1987, all those years of hard work and training with the National Guard meant nothing to the Regular Army.

A few days later, I stepped into the Marine Corps recruiter's office, sat down, and agreed to take an Armed Services Vocational Aptitude Battery test (ASVAB).

The ASVAB is a standardized test used by the U.S. military to assess a candidate's qualifications for enlistment and determine their suitability for specific military jobs, known as Military Occupational Specialties (MOS). It is designed to measure strengths, weaknesses, and potential for success in military. I qualified for a bonus and training as an electrician and Military Police. I declined both and requested a contract for the position of infantryman, 0311.

Usually, the recruiters place their candidates to feed the needs of the Marine Corps. That day, they needed grunts. My recruiter leaned back as if he held some closely guarded and dreadful secret, but I didn’t care. That slogan pierced the monotony of my life. It was an invitation to step into the extraordinary, to prove myself among the elite. I saw a way out of stagnation and a chance to be part of something bigger than myself. The Marine Corps was not just calling me. It was daring me to rise to the challenge

We Don't Entertain Stupidity

August 2005 in Ramadi brought a subtle shift in the rhythm of battle operations. Following several successful engagements against insurgents, the pace of sniper missions slowed, creating a noticeable lull. However, Shadow 1 remained active, consistently supporting Able Company in their missions. This collaboration often involved temporarily reassigning my team members to provide additional security for Able Co, which also offered an opportunity to build camaraderie and strengthen bonds between our units.

During that time, the 2/23rd STA PLT from the Marine Corps Reserve unit from Ohio was wiped out. Twenty men died violent deaths with very little shooting. Most of them were carved up. I was briefed that a team had recovered their bodies. A Marine sniper attached to one of the teams had disappeared. They finally found him mutilated. That's how the insurgents were fighting us. We were rats in a maze, trying to kill each other.

The images of that scene haunted me. Even now, twenty years later, I can still see it in full detail. Being a warrior comes with a cost so deep that most can't speak of it.

During our missions, team members like Bennett and SPC RT stood out, earning respect and appreciation for their ability to integrate seamlessly with the others. Their personable nature and skill at connecting with these teams showed the cooperative spirit that defined this period. This added depth to our work and reinforced the unity crucial for facing the challenges ahead.

SPC JY's silent nature presented a challenge for the Shadow Team. He was a great sniper, organized and acutely aware of his surroundings, so I kept him with me. Whenever we went out, I brought SPC JY. As time progressed, he would be my sounding board for ideas to improve our performance. His pet peeve was the weight distribution of his ruck. He did everything possible to lighten his kit without compromising the mission. I suppose it was also a way of killing time.

On August 23, I was lying in my bunk when a horrific explosion shook the building. I ran outside and nearly collided with another Soldier. His face betrayed his horror and surprise.

"They hit OP Hotel with a VBIED!" (Vehicle-borne IED). I ran outside and looked west towards the city where a mushroom cloud rose above the horizon.

In addition to being one of the highest buildings in the region, the OP Hotel was a four-story structure that offered blanket views of an area packed with insurgents. American Soldiers stationed inside the OP Hotel often claimed that insurgents were viewing them from a distance using binoculars, mobile phones, and video cameras. They were also hiding in vehicles and peeping around corners.

OP Hotel symbolized the chaos unfolding in Ramadi. Bullet holes and scorched streaks marred the exterior walls. The building was beaten but unbroken. Inside, especially the third floor, every inch had been adapted for survival and combat. A platoon from the 2/69 (2nd Battalion, 69th Regiment,3rd Infantry Division) had fortified the place to withstand attacks from any direction, converting rooms into makeshift bunkers and positioning heavy weapons to cover all approaches.

Sandbags surrounded MK19 grenade launchers and heavy mounted machine guns, ready to unleash a storm of fire on any enemy that dared advance. Soldiers manned observation points in every corner of the building, their binoculars scanning the city's dusty streets with vigilance honed by days, weeks, and months of unbroken tension. They sat there, eyes sharp, seven days a week, twenty-four hours a day, surrounded by stacks of ammo boxes and a floor littered in spent casings.

OP Hotel was a stronghold, where silence and calm were merely the space between barrages. They watched, waited, and readied themselves, each man attuned to the slightest movement in Ramadi's war-torn streets below.

"OP Hotel was a stronghold,
where silence where silence and calm were
merely the space between barrages."

The building's front entrance faced south, looking directly into the dense, maze-like neighborhood of Mula'ab, just north of the key supply route known as MSR Michigan. When the attack began, Al-Qaeda in Iraq (AQI) insurgents unleashed a relentless wave of gunfire on OP Hotel, targeting Soldiers of the 2/69 and Iraqi Army forces. The initial assault was a calculated distraction, forcing the Soldiers to hunker down, each man scrambling to avoid the lethal rounds that tore into the building's facade. Every corner, every window was under siege. The air vibrated with the sounds of ricochets and shouted orders.

While the Soldiers were pinned down, a dump truck heavy with explosives hurtled toward the southern wall of OP Hotel. Near the line of Jersey barriers, it erupted into a massive ball of orange flame and black smoke. The shockwave rocked the entire structure, sending debris flying in every direction.

The blast left everyone inside dazed and concussed. Two Soldiers were critically injured. The attack failed to breach OP Hotel, but it left a lasting scar on everyone within its battered walls, a reminder that death was a single breath away.

I expected many casualties and assumed they would be requesting assistance. When I ran to the FMJ building, SGT F confirmed a VBIED had hit OP Hotel, and the platoon had suffered casualties.

We were on standby to roll out. I returned to my building and grabbed my kit and mission essential equipment. Four hours later, we rolled out—Shadow 3 plus me. When we arrived, I noticed the cabin of a truck lying burnt on top of a barrier. Directly beneath the cabin was a large charcoal brick. As I got closer, I realized I was looking at a torso with no head and no clothes. Later, we would learn it was the driver's remains.

I took a position on the third floor, looking south into Mula'ab. At my 2 o'clock, a man came into view. He wore a black buttoned short-sleeve shirt, black trousers, and black shoes. He spoke on a cell phone, observed the building, and casually walked away. My gut told me he was the enemy, but he did not meet ROE requirements to kill him. I was not allowed to shoot him, although he seemed to be conducting a BDA (Battle Damage Assessment).

Later, I saw an insurgent video confiscated in the market, and the same man jumped from the passenger side of the truck that carried the explosives. *I could've killed him, and I let him get away. That will never happen again!*

After twenty-four hours in the OP Hotel, we returned to base (RTB). I considered if we, as snipers, might have spotted signs of a pending attack. Over the following days, I pondered this question. Finally, one morning, I decided to approach SGT SH, Shadow 1's team leader.

As the weeks passed, I assessed the Shadow team leaders and focused on SGT SH. He was about my height and stocky, with short brown hair, and perfect teeth that showed when he smiled. He was articulate and assertive. He delegated authority and was well-practiced in simplicity, including how he lived inside his quarters. Overall, SGT SH impressed me.

We had an erasure board with the teams labeled from top to bottom. Next to the names was another column indicating pending missions and the teams' current status. As I passed the board, I noticed Shadow 1 always had missions pending and was constantly outside the wire collaborating with Able Company.

With a dose of courage, I entered FMJ Sniper's quarters and found SGT SH lying on his cot, reading a book. I waited until he took a break, but he kept on reading. I would walk by, hoping he would stop and say hello, but the book absorbed him. Finally, I stood near his cot. After several minutes, he looked up. "What's up, man?"

I immediately responded, "I don't mean to impose on you, but I have a question." SGT SH did not move or discourage me, so I said, "How does Shadow 1 go on so many missions?"

Apparently, it was a loaded question. SGT SH did not comment on the performance and initiative of Shadow 2 or 3. He said, "Harry, you drum up your own business. Figure out where someone may need you or come up with a plan. Pass it on to SSG Gilliand. If he approves, he will let you go. More often than not, as long as you properly execute coordinating instructions, he will not interfere."

"That's it?"

"Yes, it's that simple.

Seeing we were having a discussion, I pressed forward. "I have been giving the OP Hotel attack some thought. If snipers had been present, we might have been able to catch something unusual. I'm not sure what, but I'm certain something would have presented itself with our optics. I think we should start spending time at some of these OPs (Observation Post) and ECPs (Entry Control Point)."

With his hands crossed in front of him, he replied. "Hmm, well, we don't entertain stupidity here," and went back to his book.

I was taken aback by his remark. I felt foolish and took my cue to leave. Heading back to my building, my eighteen years of police training in critical thinking kicked in. I had worked in the Special Victims Unit, where they taught us that every situation has indicators. Back then, my job was to identify the signs and properly assess them. The same was true here. I knew I was on to something. Snipers had the training and equipment to spot indicators that regular Soldiers might miss.

In the following days, I learned more about OP Hotel and its significance. I decided to move forward and act on my ideas, regardless of SGT SH's comments. I knew he meant well, but I was convinced we were missing opportunities. After considering everything, from mission preparation to rehearsals, I approached LT WH. He was tall, African American, well-spoken, and a little stressed. I offered my services, and surprisingly, he agreed. He gave me a time and a location.

I couldn't believe how simple it was. On the first night at OP Hotel, I was uncertain of our role and ability to provide services. I told the team we would have our own fire watch to avoid being caught by surprise.

Before dawn, I was lying on a cot, more or less asleep, when a distant explosion shook the building. In a flash, I rolled out of the cot and onto the floor. Getting my bearings, I heard my team's laughter in the darkness. Embarrassed, I laughed, too.

In the three days we spent there, we killed three insurgents. When I returned to base, word had already reached Shadow 6 of our performance. I did not have to report to anyone, and I was glad to get much-needed rest. I was pleased with how things went. More importantly, we had won the confidence of Able Company 2/69. This was a relief because the Regular Army could send Shadow 4 packing. We were National Guard, and we couldn't forget that.

I had maturity from my Marine training and police training, plus years of working on the streets. This made me a decisive combat leader with a different mindset than most of the others, including those in leadership. I had the confidence to speak up, but also the humility to know that the best plans can end up in disaster. Shit happens.

I felt like a scientist trying out my theories, examining the results, and refining them for the next try.

Later in the day, I went to SSG Gilliland, and we discussed our mission at OP Hotel. After that, Shadow teams rotated out of OP Hotel for the duration of the 2/69's deployment.

WHO THE FUCK KNOWS?

Afterward, I decided to go to Entry Control Point 5, a platoon-sized control point covering a bridge and occupying three dwellings. Taking the same approach, I managed to get us assigned to ECP 5 as snipers, with the understanding that we managed ourselves and that we were not on guard duty but a force multiplier, taken right out of the manual.

We were well received, but it was a formal reception. The NCOIC had already experienced a well-coordinated and complex attack by the insurgents. Although everyone was slightly on edge, they radiated confidence, and they richly deserved it. They held up well against an attack coming from three different sides. They must have felt like the whole town had risen up against them.

The 2/69 Armor (AR) showed up with its Bradley Fighting Vehicles (BFV). They roared down the street and saw the attack was coming from the mosque. Aiming its Bushmaster, the BFV unleashed a hornet's nest upon the minaret, the tower where they call believers to prayer.

Immediately, every military vehicle turned on the minaret, firing to stop the threat. Moving closer to ECP 5, the BFV abruptly stopped. The back door dropped, and 2/69 Soldiers poured out onto the death ground. Those Soldiers quickly fought their way to the ECP, where they continued to fight the enemy until the insurgents retreated into the populace and disbanded. ECP 5 held its position.

An Animal Company Platoon manned ECP 5. Their PSG (Platoon Sergeant) was a former Marine known as "Gunny," a title used for Gunnery Sergeants in the Marine Corps. With that in mind, I approached him, my tone friendly. "Hey, Gunny, we came out to offer some assistance."

Annoyed, he replied, "Specialist Brown will show you your quarters. I know you need to follow a different ROE." (Rules of Engagement)

My police voice came online. I interrupted him. "Sergeant, we follow the same ROE as everyone else. We do not ask permission to engage the enemy. My guys are well-trained and know when to take the shot." I wanted him to understand I did not care for his rank, nor would I change our level of performance to appease his sense of righteousness. This must have been strange to him because I was an E5, which means three stripes (three up), while he was an E7, meaning three stripes and two rockers (three up and two down). He also knew I was a former FMF Marine. One does not get in the path of a Marine with a task and purpose. That only generates pain.

"One does not get in the path of a Marine with a task and purpose. That only generates pain."

That night, we stayed in a house with cemented walls and large rooms, which added surrealism to the situation. Inside, Soldiers were doing the typical thing, talking about home and complaining about their predicament. Infantrymen have done this since the days of antiquity. Some things will never change.

We went to sleep. The next morning, SPC JY and I walked around in boonie hats to assess the location and possibly build a hide. While up on the roof, Gunny approached us and ordered us to wear our helmets.

"Negative, Sergeant," I said. "We don't wear helmets, and we do not wear the issued armor." Unbeknownst to anyone, I had authorized carrier plates for the team. The plate carrier kit was less bulky and cumbersome than the issued gear. I am convinced someone's mom designed the Army's issued Improved Outer Tactical Vest thinking about safety instead of fighting the enemy.

I wore an olive-green baseball cap with the symbol of the eagle globe and anchor subdued. I'm sure that irked him, considering I represented the Army. Later, we would continue to dress according to the mission. It meant removing the Polar Tec black fleece, adopting a hunter-green fleece, and not wearing anything black on my head. I had found a wool beige cap with a scarf to match. As long as no one who mattered complained, I was quietly blessed with my free thinking.

Gunny walked away with his thumbs hooked under his armor. I remarked to my team, "Don't worry about him. He ain't shit, and he knows it."

As the day continued, we kept busy walking to random positions and talking to different Soldiers, trying to understand the purpose of ECP (Entry Control Point) and how it functioned. I wanted to know what the enemy had seen and perhaps understand what encouraged them to attack.

An American platoon from Animal Company (PA ARNG) manned ECP 5. Another platoon relieved them every three days. The town of Sophia knew when this exchange occurred because a convoy of HMMWVs would suddenly show up sometime in the day. Then, the convoy would depart using a different route at an undetermined time. The Soldiers occupied three homes along a narrow canal. The distance between the closest and farthest residences was approximately 400 meters. The residence in the center was unoccupied.

I wanted to insert my team into a random residence, but that night we could not, due to manpower issues. After a couple of days, we were moved back to base with the outgoing platoon. Although we were not well received, I knew we would return.

Upon RTB, I ate chow and decided to walk over to FMJ. On the board, two of the teams were inside the wire. As usual, Shadow 1 was operating in Ramadi. Next to Shadow 4, in the two columns covering pending missions and current missions, someone had scrawled in black Sharpie, WHO THE FUCK KNOWS.

If I were among friends, I would have laughed. As it was, I could have gotten mad, but that was not the proper response. I turned around and walked out. I was concerned my team might be alienated because of my ambitious efforts.

I was running my team from mission to mission, and some of my ideas appeared to be working. I started to see more Shadow team members on the firing range, so the unit as a whole was evolving. Shadow 4 was operating out of OP Hotel, and I began speaking up, whether we were asked to speak or not. We were piling up an enemy body count, which bothered some team leaders. Shadow 1 never made any remarks, but I could sense bitterness from the other two team leaders. We were from the National Guard, and the Regular Army didn't like to be shown up. Rumors began circulating about Shadow 4, and I was concerned that we would eventually run into blowback. I didn't want my men to go through tough times because of my ambitions. We were already going through a lot just by being there.

Later, I sought out SSG Gilliland to talk through my role and my thoughts on the day's events. We went over what I had learned—both the insights that could strengthen Shadow 4 and the habits that might put us at risk. On many points, we were aligned. His main advice was clear: keep him informed and always discuss plans ahead of time, ensuring every move was deliberate and well-considered. I respected his perspective and his decisiveness. He had a sense of responsibility in his approach that I trusted without question. Loyalty to him was not forced or even a choice. It felt like a natural extension of the trust and respect he earned through his leadership.

After a few hours of sleep, I got up and walked towards Animal Company to show my face and let them know I was alive. By this time, Shadow 4 had become a separate entity from the PA ARNG. My CPT and 1SG were great. I always greeted them. Informally, I updated my commander on our missions and the results. My Soldiers were still Animal Company, known and well-loved. In this manner, the CPT allowed me to take the initiative and assert my role as a small-unit tactics leader with debriefings after missions that I found helpful.

I had left the Captain's office with a purpose, planning my team's next sniper mission, already running through the details in my head.

Walking back toward the barracks, lost in thought, the thunderous impact of an incoming mortar round struck the entrance of 2/69 Battalion HQ.

CHAPTER 5
INTO THE CITY

Our view of Eastern Ramadi, MSR Michigan, looking west from OP Hotel August 2005

Jolted to the ground, the blast rang in my ears. I staggered to my feet, instincts kicking in before the shock had faded. I ran toward the entrance, where Soldiers and civilian interpreters lay scattered across the ground. Blood smeared the dirt. The air was thick with dust and the metallic scent of explosives.

I dropped to my knees to check the wounded, scanning faces until I spotted Sam, a friend and a private contractor hired to assist American forces as a translator. My heart twisted as I saw Sam lying there, his hand weakly pressed to his forehead. I knelt beside him and grabbed his hand, steadying myself as much as him. "Sam, you're going to be fine. You're just banged up a bit," I said, mustering up strength in my voice I didn't fully feel.

Sam's eyes opened, and for a moment, he was there with me, his gaze clear.

"Harry, please—find my glasses. I need my glasses." His voice was soft, yet urgent, as if those frames were his only tether to normalcy. I nodded, taking note of his face and arms where small puncture wounds from shrapnel had left their mark. For a brief second, I thought he would be all right. We lifted him onto a stretcher, getting him inside for first aid. I promised I'd find his glasses.

Outside, I combed through the debris, searching until I spotted them—a twisted pair of frames lying in the dust, smeared with blood. I wiped them clean with my shirt, trying to straighten the bent metal, bringing them back to something like their original shape. When I found Sam again, he was being loaded into the casualty vehicle. I slipped the glasses gently onto his face, and he looked up at me, the smallest smile of recognition breaking through his pain. The doors closed, and the vehicle pulled away, carrying him off into the fading light.

After dinner, I walked back to FMJ. Soldiers were removing Sam's belongings and clearing out his space. Sam had died from his wounds. Losing him hit me hard. I still feel his loss to this day.

We'll Wolfpack the Street

While SSG James Gilliland was away on much-deserved leave, and SGT SH was back in CONUS (Continental United States) escorting a fallen brother. By default, SGT FN was in charge. Tall and fair-skinned with Irish freckles, SGT FN had the heavy Georgia accent I heard quite often. He was about thirty years old and a former Sheriff Deputy. I knew that most of Shadow team had lost respect for him. Still, SGT FN went out of his way to extend a hand in friendship. I had mixed emotions about that. I wasn't sure how to take him.

In the absence of SGT SH, SPC PR was temporarily assigned as a team leader, and we were having lunch one day to discuss ops. Battle rhythm had decreased under SGT FN's rules. The missions were also low risk. Once, we escorted a Human Intelligence collector. On another mission, we attempted to infiltrate a town to coordinate a jet dropping an LMAV missile on a suspected enemy sniper hide, but we did not have JTAC (Joint Terminal Attack Controllers). I couldn't understand how we could accomplish the mission without them. While enroute, the mission was aborted.

After a couple of aborted missions, SGT FN decided to put our skills to use. As it turned out, AQI had heavily mined a road with IEDs.

We tried to walk parallel to the road using a dry narrow irrigation route, but as reports of IEDs came in, we decided to abort the mission. Along with my M16A4, I was carrying a suppressed M24 strapped over my assault pack. As we walked, I could not get it into a comfortable position. Every few meters, I kept adjusting the strap. According to the guy behind me, I looked like I was wrestling with a monkey on my back.

After returning to FMJ, some of the team members expressed frustration with the nature of these missions. I decided to take advantage of it and proposed several plans. One that we all liked involved moving out into the industrial area and setting up a hide while allowing a subtle compromise to our position. Then, we would select another hide and covertly move into it, keeping the original hide under surveillance and waiting to see what happened. The idea required reviewing the record of all Significant Activities (SIGACTS) in our area of operations and finding a road we had not worked on in a while, one that would likely have IEDs now or in the recent past. Insurgents avoided digging because of the fear of being killed. They preferred to simply reseed the hole where an IED had been.

One day, a company from 2/69 AR was traveling south on No Name Rd. when one of the armored personnel carriers, an M113, struck an IED. Sadly, a 3rd ID Soldier was killed. I figured now was the time to say something about my ideas to kill the insurgents. Between missions, I had been jumping on Dagger Route Clearance Missions and noting where we were collecting, diffusing, or blowing up the IEDs.

An area rich with IEDs was the section of the city known as Mula'ab. Up to this point, Shadow team had never ventured directly into the city with the exception of SGT SH's team who executed an overwatch mission near the bridge above Mula'ab. After joining multiple patrols inside Mula'ab with the engineers and careful analysis of their operations, I had an idea. I proceeded to FMJ, Shadow's hooch where I found SGT FN, SGT LL, SPC PR and several other members.

I pitched my idea to SGT FN while SGT LL stood nearby, listening. SGT LL looked like a Soldier—muscular and fit with a dynamic personality. I had heard the story of SGT LL's heroism when his team came upon an enemy sniper. LL decided he would go in and take him. Armed with only a pistol, he entered the hide, overcame the sniper, and brought him back alive for questioning. He went into that closed environment knowing the man was trained to kill him whether with a weapon or hand-to-hand, and SGT LL prevailed.

I respected him for that, but his loyalty to SGT FN brought his judgment and resolve into question. Regardless, I explained to them my analysis based on intelligence reports and excursions with TF Dagger into Mula'ab. I suggested that we needed to enter their neighborhood and disrupt their efforts, so enemy teams and equipment would move into the neighborhoods north of MSR Michigan, and we could hit them in their own backyard.

The moment I suggested Mula'ab, I could see by their facial expressions I was going nowhere on the idea. Initially, I suggested a foot recon to be followed up later by a sniper mission. As I continued with the proposal, they shook their heads, refusing to maintain an open mind and listen. I thought, maybe if I ask if one of them to come along, we could share responsibilities.

When I did, SGT FN said, "OOOOWeee, no way."

"Why not?" I responded, trying to keep cool.

"Because it's suicide," he said.

"No buildings afford a shooting opportunity," SGT LL added, with his hands on his hips.

I countered, "We don't know that. We haven't done a proper recon. From the satellite imagery, there appear to be some good homes we can seize."

"No. Forget it," SGT FN said.

SGT LL closed in. "Why don't you let us do the planning?" It was more of a statement than a question.

I was losing my cool. I shot SGT LL a shut-the-fuck-up look, and spoke to SGT FN, "I hear what you are saying, but no one has truly ventured into Mula'ab. Our biggest advantage is the element of surprise. They do not expect us inside Mula'ab. We have several points to exfil if problems arise. We have OP Hotel to our north and the entire Camp Corregidor to our east. Delta Co has tanks! I don't see a problem." I was getting desperate.

SGT FN's voice rose, "No. I'm not going to repeat myself."

My voice rose to match his. We were all the same pay grade and rank, E5s. "You are wrong. I will lead the team on the recon and the mission itself. There are guys here that will go!"

SGT FN exploded. "I'm in charge. This mission will not happen." He waved his hand. "Listen. Get the fuck out of my house. Don't raise your voice in my house. Get out!"

"You are being too cautious. These other missions are pointless, they should..."

"GET OUT! Don't talk to me about other missions! Get the FUCK OUT!"

As I faced SGT FN, SGT LL stood to my right, his elbows slightly bent, feet wide. I was willing to fight them, but I figured I would lose. The pummeling didn't bother me but, worst-case scenario, I could be dismissed from the section. That was not worth an altercation.

I left. I was confused by their reluctance and lack of fighting spirit. I felt humiliated and powerless, but my greatest worry was that my guys would suffer at their hands because of my actions. Bennett and SPC RT had already complained about SGT LL's incompetence. In one case, they were compromised and nearly killed by the Iraqi Army because of his actions.

SGT FN didn't trust his own team.

After the altercation, I avoided both SGT FN and my team because I was embarrassed by my outburst inside the hooch. I also wanted things to calm down.

Situations like this taught me to be ruthless in my own selection for leadership positions. If someone comes in with red flags, their weaknesses will jeopardize the unit and their missions. You cannot train red flags out of them or influence them to improve. Better to fire them fast and early to make room for someone more suited to the position.

In the evening on the following day, I went back and found SGT FN outside the door. I wanted to apologize, but I could not. I felt I did nothing wrong.

After a moment of awkwardness, SGT FN spoke in a calm voice. His Georgia accent was heavy and sincere. "Harry (he pronounced it *Hairy*), I'm sorry about the other day. I lost my temper. You are a good guy, but we have experience. I expect you to respect my orders."

I extended my hand, and I apologized for raising my voice. Honestly, although I disagreed with him, I respected his ability to apologize. I could live with it.

We went inside together and resumed our normal routine. We joked around and discussed some unrelated topics. After a brief time, I left. As I walked away in the darkness, I thought about formulating another plan. So, I went back to my books and began to read *The Ultimate Sniper* by MAJ John Plaster. As I turned the pages, an idea began to develop.

The idea that came to me was called wolf packing, using multiple teams over a long distance, covering the same area. Early on, I realized the enemy was choosing targets where there were no American observers or where no American patrol had been for several weeks. So, the first step would be to select a street where we had been some weeks before, and where they wouldn't expect us to show up again. Most importantly, after we had collected an IED in that area diffused by TF Dagger.

I spoke to SPC PR, to Able Co., and to my team. Shadow 4 and a security team from Able Company would occupy a preselected hide for twenty-four hours. We would be the second team going in, since I strongly believed that during the second twenty-four hours, we would engage an IED team or witness an event that qualified for engagement in accordance with ROE.

SPC PR was interested. We pored over the maps together. We checked our INSUMP (intelligence summary report) reports, and we discussed tactics as well as the concept of operations. We also war-gamed, pointing out critical areas of the operations, basically a what-if session. At the end, we agreed that our biggest hurdle would be SGT FN. We pulled out the calendar, marked the day for SSG Gilliland's arrival, and waited.

I was concerned SSG Gilliland would become angry and fire me from the team after discovering I had openly argued with a Regular Army NCO appointed by the section leader. When nothing happened, I decided to pitch my idea. SSG Gilliland approved the mission. Apparently, SPC PR had already discussed the mission with him, and PR would lead the second team. Smiling to myself, I agreed, thinking, *Damn, he's going to see something. He knows we put this together, so this should buy me some credibility.*

November 6, 2005, at 0100, was the first time I generated an entire mission from concept to execution, led entirely by a National Guard NCO and performed by a National Guard sniper team, supported by the Regular Army.

After dinner, I grabbed some sleep until SPC JY's voice brought me around. "Sergeant Martinez, you awake?"

"Yeah. What time is it?" I muttered, my voice heavy with fatigue. We had a big day ahead of us, and I wanted every second of shut-eye I could get.

"2345 hrs. Are you packing any snivel gear?" JY had begun to open up and talk more. Slowly but surely, we were finally connecting. The kid was coming to terms with a hard truth: survival in this place depended on trusting each other to get through alive. I made a point to listen to him, to take his questions and actions seriously.

Despite the almost-twenty-year gap in our ages and experience, JY contributed something valuable to our collective success. His perspective helped ensure victory on what I called the *death ground.* That term felt truer to me than battleground. While battle hinted at strategy and heroics, death ground lay bare the unflinching reality of slaughter and brutality. That's why we were here.

In Ramadi, every decision, every connection, meant the difference between life and death. SPC JY was learning that truth, and it would shape him as it had shaped me.

I swung my feet out of my rack and placed them on the cold floor. Everything was covered in desert dust. You couldn't escape it, a tan film of talc powder that we lovingly called moon dust. I said, "I'm not bringing my sleeping bag. It's only a twenty-four-hour mission. I think I can manage the cold for a few hours."

Mornings and evenings brought a distinct chill, with lows dipping into the 50s°F (10–15°C), a stark contrast to the sweltering days of summer. These cooler nights were often accompanied by a biting wind that coated everything in a fine, gritty layer.

November also marked the onset of Iraq's brief rainy season. While rainfall was still rare, occasional light showers or overcast skies hinted at approaching winter. The rain turned the dusty streets into slick, muddy paths, complicating movement and operations. The weather mirrored the volatile environment—unpredictable, unforgiving, and always demanding adaptation.

SPC JY informed me that he had packed his polypro top and wore his ninja suit (a wick-away nylon top and bottom) underneath his uniform. I thought about telling him to take it off. My experience in the Light Infantry Fleet Marine Force told me it's wise to ruck in your BDUs while wearing a t-shirt despite the chill in the air. JY was a National Guardsman. He had trained as an Armor crewman where tanks drove everywhere. I decided that JY would have to find out the hard way about what to wear. Some things were meant to be experienced.

We were headed to a dangerous stretch of road nicknamed No Name Rd. by American forces. This road was saturated with IEDs. We were going to occupy a house overlooking the road, wait for insurgents, and kill them.

The uncertainty of whether we'd make it back alive wasn't a passing thought—it lived in the back of our minds every step we took. AQI had its claws in the local population, twisting fear into obedience. They recruited, coerced, and manipulated civilians until you couldn't tell who was trying to survive and who wanted you dead. This was more than a fight with rifles and roadside bombs—it was a war of trust, of perception, where the line between friend and enemy disappeared. Every face, every movement, every silence carried weight. Our mission went beyond hunting the enemy. We had to move through a place where the rules kept changing, where an entire community was caught in the grip of something darker than we could see.

We had not worked on that street for the past six weeks, and CF (Coalition Force) units had already suffered human loss there. Insurgents still believed it was a "Kill All Americans" national holiday. According to news reports, American Forces in Iraq had just survived the third-deadliest month of war since March 2003. Hajji was in for a surprise. The party was over, and it was time to clean house.

JY exited my room. My rifles hung on the wall, as did my rucksack. Next to that were different pieces of essential gear. My wall was decorated with quotes and pictures of my family. Each item was a factor in my survival, both spiritually and physically. I got dressed and methodically donned my gear. My weight in my birthday suit was a lean 158 lbs. After adding my tools of war, I weighed somewhere between 200 and 220 lbs. On some occasions, I tipped the scales at 230 lbs.

I found JY waiting for me in the hallway, a quiet but meaningful gesture of teamwork. Without exchanging words, we stepped out onto the road. Walking together as a group was a small ritual that carried weight. I hated the moments when we had to part ways. Together, we were ready to face whatever the night might bring.

Tonight was different. I was leading these young men, full of life and potential, straight into harm's way. The weight of that responsibility settled heavily on my shoulders.

"JY, go down to the front gate and meet with the other guys there. I have to go to BN (Battalion) and inform TOC that we are ready to head out."

JY responded, but I didn't hear him. I looked up into the sky. A partial moon. *That's good, and that isn't good.* Thus far, the insurgents still lacked night vision equipment. In Ramadi, we owned the night, and I was looking for a fight.

I proceeded to the TOC and coordinated all the necessary details with the Battalion Operation (BN OPS) staff inside. To them, I was just another patrol in the city. I briefly wondered: *Will I come back tomorrow? Will my team? Am I full of myself? Will my plan work?*

I shook those thoughts from my head and quietly whispered a prayer that I had written on the wall inside my living space. It's a modification of the widely circulated poem, "A Police Officer's Prayer" (author unknown), but it also applies to war. I repeated it as part of my routine while preparing for a mission. I said it as I made my way to join Shadow 4.

> Lord, I ask for courage.
> Courage to face and conquer
> My own fears.
> Courage to take me where
> Others will not go.
> I ask for strength.
> Strength of body to protect
> Others.
> I ask for dedication.
> Dedication to my country
> To keep safe.
> Give me, Lord, concern,
> For those who trust me.
> And compassion for those
> Who need me.
> And please, Lord…
> Through it all… be at my side.

Standing at the gate was SPC RT, who was studying to be a mathematician. He was tall, confident, and looking forward to reuniting with his fiancée. SPC Joey Bennett was studying to be a pharmacist. He was also tall with a mop of curly light-brown hair and a lethal fixation on deadly booms. JY was there, silent and vigilant as ever.

We were Shadow 1, snipers who struck fear into the heart of the insurgency. We hid in the shadows and used every advantage to perform our lethal task and walk away, victors of the death ground.

We were intelligent, assertive, self-reliant, bold, and sometimes wholly arrogant. We had all volunteered, not once but several times. We volunteered to join the Army National Guard. We volunteered for Combat Arms, and we volunteered to become snipers.

We made last-minute checks, locked and loaded our weapons, and adjusted our gear for the one-hundredth time. As we crossed the street to the COP (Combat Out Post), we harassed each other good-naturedly. Not surprisingly, most of it was directed at me. Since my arrival in this sandbox, I had inadvertently created a style for rolling my ankle and falling. Believe me when I say they were spectacular falls, the kind that make a stuntman's career.

The road was clear, and the four of us meandered into COP, another small FOB lacking all the pleasant luxuries of life the American way took for granted and providing us with a deeper definition for shithole.

At COP, we linked up with three members of Able Co.—one NCO (Non-Commissioned Officer) and two Squad Designated Marksman, the newest role the Army developed in the Infantry. A great concept, but like the sniper, it caused a stir and created confusion for the ignorant and the hapless that sometimes filled the ranks of leadership. Right then, that was not my concern. We quickly shared information about the mission. We rechecked our gear, reviewed the map, and verbally rehearsed our roles in the mission.

JY removed his ninja suit. He was hot. RT and Bennett shared last-minute stories, while the Able element went over their gear and eyed us suspiciously. Able Company had already established itself as a fighting unit under competent leadership. Although it had suffered the most casualties in the Battalion, Able Co. was still in the fight.

The Regular Army and the National Guard had a tumultuous relationship, often on the brink of open hostility. The Regular Army did the job 24/7. We were part-timers, weekend warriors, and wannabes.

I didn't blame them. I would have done the same. By now, Shadow 4 had a positive image, but we were still the National Guard. And no matter what, we would always have to prove ourselves.

0200. Time to execute the mission. No more joking. Final adjustments and another prayer in our own words: *Lord, don't let me fuck up. Worse, don't let me fuck up and live!*

I took point. By now, I was familiar with the streets and felt confident I could lead my team to the location without feeling overwhelmed. The house was located nearly 500 meters from the front gate. Under normal circumstances, it was less than a five-minute walk. In Ramadi, in combat gear at a combat patrol pace, walking in a roundabout manner to the hide, we arrived a full thirty minutes later.

We climbed over the wall and made our way to the front door. We had been to the hide in the past. I was certain the house owner would allow us to use his house. His child had been killed as a result of an IED, and he wanted all insurgents dead.

Although he invited us in, we cleared the house. Once we were satisfied. I told the owner we were here to kill insurgents, and we were going to be attacked. I told him to leave. He insisted on staying and supporting our efforts. I told him I was concerned for the safety of his family. Could he find shelter elsewhere? He paused, giving it serious thought.

As he pondered, I added, "It's going to be bad."

"All right. We will leave." He gathered his family and left late at night.

This was the first of many changes I would make in Shadow Operations. Normally, we would keep the family isolated somewhere in the house with security posted to watch them. As a leader, I assumed the risk and trusted the family that they would not compromise my mission. I allowed them to leave. This turned out to be a solid decision.

This kind of flexibility allows leaders to assume risk and encourages an aggressive and commonsense approach as each situation presented itself. Officers call this a mission command philosophy, and it's practiced by seasoned combat leaders.

While security set up below, Shadow 4 went upstairs. We divided into two teams, separated by a hallway. The second floor was under construction and had no roof. The windows had no glass or curtains. They were covered by a black iron grill to prevent looters from entering the premises. We set up our nets and began to put together hasty shooting platforms. A radio call was sent to the BN TOC with a SITREP.

While out on missions, I never slept at night. I waited until the sun came up and all security was in place. Only then would I feel comfortable enough to close my eyes. This night was no different. While my team slept, I implemented our rest plan and kept vigilant watch during the hours of darkness using my NODs.

Around 0600 with the sun above the horizon, I woke up JY. I was tired and slightly chilled from the morning air. I leaned in close and spoke in a low whisper, "I want you here in this room looking south. If they are going to put in an IED, it will be between now and 0700. Do not fall asleep. If you see someone doing something stupid, grab RT. The two of you kill the bastard. Do you understand?" When he nodded, I went on, "I am going downstairs to sleep. Wake me up around 0800 or 0900. It doesn't matter which. Do you have any questions?"

"No." JY never wasted words.

I went downstairs and found a good place on the floor to sleep near the staircase, which would provide me with cover and concealment from enemy fire or rockets. It was near the security team and warm, perfect for me. I closed my eyes and drifted away.

BANG!

I sat straight up. The sound came from upstairs. I grabbed my weapon and rushed upstairs. I found JY sitting with his rifle resting on his shooting platform, staring intently south. Next to him, RT had an excited look on his face. I looked at my watch. 0700. I could not believe the timing.

I crouched next to JY, who remained focused on the street. "Talk to me, JY. What did you see?"

With his eyes on the street, he told me a MAM (military-aged male) had been seen traveling north on No Name Rd. After passing a mound of construction dirt, the MAM turned around and dug into the dirt mound.

At this point, JY called RT over and instructed him to be ready to kill the MAM. After digging into the soft dirt, the MAM removed his backpack and placed a bagged device into the hole. After a brief time, the MAM put his backpack on and mounted his bicycle, traveling south.

During this time, JY brought his weapon to his shoulder, calculated the distance to 100 meters, and counted down from three. He and RT shot simultaneously. The man fell over with his legs still wrapped around the bicycle frame.

The body still lay in the same position.

"Good job, JY. You killed him. Bennett calls it in." I grabbed my camera and took a picture of the dead insurgent. (Later, I learned he was wounded and laying very still.) In that moment, I decided to continue the mission and kill those who were coming to recover their comrade.

I found something to sit on to wait out the situation. On other missions, we would have broken down and moved from the house back to Camp Corregidor.

Someone asked, "Are we leaving?" Everyone's actions indicated that they thought the mission was nearly over.

I felt it was only half over. We had no signs that we had been compromised. "No, we are staying. Get back to your positions and be ready for anything."

Nearly half an hour passed. One of the security guys came upstairs, bringing me a cup of Iraqi chai tea. Before I could take it from him, someone called out, "SGT Martinez, we got a vehicle turning around."

I was watching another vehicle through my binoculars. This one stopped, and a man exited from the rear passenger door to have a conversation with the "dead man" on the ground. Suddenly, the "dead man" pointed south, as though pointing out where the shot came from. He was wrong. We were to his north.

I jumped up and ran to the window, shouting, "Kill them all!"

CHAPTER 6

AIYA

3rd ID Able Company, Team leader handing his Soldier a grenade to repel the enemy's assault, No Name St. Ramadi Iraq November 2005.

Small arms fire came at us from their direction but did not impact our position. We had positive identification (PID) and met the requirements for rules of engagement (ROE).

The air inside on the second floor was thick with tension and cordite as RT and Bennett fired relentlessly from the other room. Every corner of the firefight bristled with life-or-death decisions. I steadied my breathing, selecting targets and trying to maintain calmness with a breathing technique I learned on my SWAT team.

Through the bars covering the window, I saw the first vehicle abandon its injured, leaving the "dead man" and his would-be rescuer exposed. Flashes from the second vehicle betrayed the muzzle fire inside. I locked my aim onto the windshield, watching my rounds tightly crater the glass.

Bennett's voice cut through the chaos. "SGT Martinez, can I use my hand grenades?"

"THROW THEM! THROW THEM!" I barked back, barely letting him finish. The first grenade landed wide across the road, but the second exploded dead center, a shockwave that seemed to freeze time. The enemy vehicle screeched to a halt, skidding dangerously close to our courtyard gate.

"They're dismounting!" The shout galvanized me into action. Grabbing a grenade, I flipped it upside down, thumbed the clip, pulled the pin, and lobbed it high, aiming for the insurgents behind the wall. The explosion reverberated far too close—it had landed short, inside the courtyard.

Without hesitation, I sprinted downstairs, finding SGT NH amidst the dust and debris, his expression dazed.

"Hey man, how you doing?" It was a ridiculous question, but the semblance of calm was all I could offer.

"Fuck, SGT, I think an RPG just hit the courtyard."

"Yeah, about that... There are bastards out there who need finishing off."

NH straightened, shaking off the confusion. "Yeah, no worries. PFC Mac, grab your gear—we're heading out."

I wasn't expecting such boldness, but there was no arguing with a man and a good plan. "Cool. I'll get back upstairs and cover you."

The firefight continued in full force as I returned to my perch. JY was exchanging fire with a figure who popped in and out from behind a wall. "Wait him out," I told JY. "He'll show himself again."

Like clockwork, the insurgent rose. JY and I fired. His body jerked back, his head snapping up before he crumpled to the ground.

Suddenly, sharp cracks rang out close to the house. At the west window, I spotted Mac coolly engaging insurgents through a narrow gap in the wall. When his rifle went silent, NH handed him a grenade. Mac launched it without hesitation, the explosion a grim punctuation to their retreat back into the building. In that moment, they embodied something raw and primal, a message delivered with unflinching savagery. You attack us, and we will destroy you.

Bennett, our radio man, was still firing as he updated me. "QRF's ramping up, but there's a delay. I've been busy shooting."

"Well, get ready. We're moving soon."

Everyone's breathing was heavy, but their movements were deliberate, checking weapons and gear, preparing for the extraction. As we regrouped downstairs, I saw the security team poised, ready to step into the chaos once more. We were bloodied but intact—a testament to our training, our bond, and our refusal to give an inch.

"Everyone okay?" I asked no one in particular.

I heard random yeses and counted seven friendlies. We were up.

The courtyard was eerily silent as I gave the order. "Okay, let's head out." A sense of dread surrounded me. We might be walking into a trap. The number one man through the gate was almost certainly a dead man if insurgents were lying in wait. I couldn't send one of my men ahead, nor could I ask the security team to do what I wasn't willing to do myself.

"I'll be number one, then RT, Bennett, and JY, followed by SGT NH's team." I stepped forward, gritting my teeth as I slowly pushed open the gate.

Stepping onto No Name Rd, the world felt heavy with heat and danger. The air shimmered with dust and smoke as the relentless Iraqi sun bore down on us. Barely five feet from the gate, I took in the grim scene. A car to my right held two lifeless bodies. On the ground was a dark, glistening trail that snaked away from the house and around the corner.

The radio crackled. One of the Shadow team members stationed nearby said, "Get back inside the house! They're still out there waiting to kill you."

With no sign of the QRF, we were exposed. I quickly signaled the patrol to fall back. I turned to my number two man, speaking low and urgent. "Let's get back inside. Take up a 360 position with your grenades and mags, and wait for QRF."

Moving swiftly, we withdrew to the house and locked down defensive positions. Minutes crawled by, stretching into an eternity. Every second dripped with tension as we scanned for any sign of movement. Finally, the distant rumble of the QRF approached from the south. We popped smoke to mark our location, expecting the Bradley Fighting Vehicles (BFVs) to roll up.

Bennett's voice cut through the static: "SGT Martinez, they're not coming down the street. They want us to move to them—IEDs still active."

"Fuck me!" I snapped, frustration boiling over. "Mount up! Let's go!"

We grabbed our gear and moved out. Again, I took point. As we hurried toward the BFVs, a man on the ground ahead caught my eye.

He was lying on his back, but then he moved. He was still alive, reaching into his jacket. His voice was clear: "Mister? Mister?"

Time stopped. His hands moved deliberately into his brown jacket, his dust-covered face and short-cropped hair vivid against the sun. Every fiber of my being screamed danger. Without hesitation, I raised my weapon and fired three rounds into his chest. He slumped back, lifeless.

Suddenly, rounds snapped past me, kicking up dust. I turned left toward the source of the fire and saw a shadowy figure on a rooftop, barely visible from the chest up. I saw no details, just a dark figure resembling a 50-meter target on a range. My training took over. I dropped to a knee, exhaled, and squeezed the trigger. One shot, then another. The figure crumpled out of sight.

At ground level, the security team was already engaging. They poured suppressive fire onto insurgent positions, moving with grim determination up a dirt mound toward the enemy's cover. The intensity of the firefight was deafening, cruelty in every direction.

With a distinct roar, the BFVs came to life. The Bushmaster 25mm cannon unleashed a torrent of hate and death, its heavy rounds ripping through enemy positions with savage precision. The ground shook with every burst, each round delivering destruction. This was the fierce, unforgiving language of war when our lives were measured by inches and seconds.

"This was the fierce, unforgiving language of war when our lives were measured by inches and seconds."

Someone shouted my name. SFC O was yelling at me to get on board. His face was red, and his mouth was wide open, like a demon unleashed by hell. "C'mon, you motherfucker, let's go!"

It sounded like a promising idea. I got up and ran as fast as I could. When I got on board, I realized I was the last one in. I never saw anyone run past me. Immediately, the BFV closed the ramp, and we were bathed in darkness. The ride was fast and furious. We came to a screeching halt at the COP.

As I dismounted, SGT NH walked up to me. He said, "I will go out with you anytime." He had a big smile on his face.

All I could say was, "Thank you."

I looked at my guys. "Good job. Get some rest, and I will report to the TOC to do the debrief."

After the debrief, I proceeded to Shadow Team's hooch. I woke up SPC PR. Their room was dark. Everyone was asleep. As he rubbed the sleep out of his eyes, I gently informed SPC PR in a whisper, so as not to disturb the others, "Mission complete. We confirmed five kills and three probables. No friendly casualties." I patted his shoulder. "Get some sleep. We'll talk later."

Aiya's House

In selecting our next hide, we used a drone, followed by a recon foot patrol to find a house that would allow us to engage as far as 800 meters into Mula'ab. When we selected a house, we discovered a wall on the second floor denying us entry into the room we needed to access.

Initially, we used sledgehammers against the wall, but so early in the morning, pounding at a wall made no sense. We would announce ourselves to the neighborhood. We left the premises, intending to return within a few weeks and warned the lady not to rebuild the wall. If she worked with us, it would be to her benefit. We would help them find employment with local Iraqi government or along CF (Coalition Forces) units.

With a steely gaze, she replied, "Why would I help Bush? He took everything from us." She would have spit on us if she could get away with it.

I turned to my team and ordered them to collapse to collapse our position.

When we arrived at Camp Corrigedor, we stopped to have a conversation at the edge of a shooting range. Some sat on the ground while others knelt, the warm sun glinting off the brass casings scattered around us. The air smelled faintly of gunpowder and the churned-up dirt from countless boots. My team formed a loose circle, leaning in as we hashed out a change in tactics. Voices rose and fell with animated debate—some scratched diagrams in the dirt, others gestured to illustrate their points. Rifles lay within arm's reach, their silent presence a reminder of the weight of the decisions we were making.

This scene wasn't only about tactics but about the conversation itself. These discussions built trust and cohesion. By inviting every team member to contribute, I aimed to instill a sense of ownership in the plan. When a Soldier feels like a stakeholder, they're not just following orders—they're thinking critically about their role and how it fits into the bigger picture. Each shared idea wove us tighter as a sniper-heavy team. Sometimes, team building is about raw, unfiltered moments where every voice matters.

"Sometimes, team building is about unfiltered moments
where every voice matters."

Sitting there, surrounded by Shadow 4, I could feel it—the quiet transformation of a group of individuals into something greater: a team that trusted one another to execute a shared purpose.

Suddenly, I had an inspiration. "I have an idea. I know it sounds crazy, but why don't we blow a fucking hole in it? Hajji is used to hearing explosions. He'll hear it and shrug it off."

In a mocking tone, I was trying my best to use an Arabic accent. "'Oh my, Omar has exploded another bomb against the infidel. Oh well.' Then he'll roll over and go back to sleep."

Everyone stared at me, confused.

"I'm serious, guys," I went on, warming to the idea. "Let's blow a hole instead of digging. We'll talk to Navy EOD and ask them how to do it, train until we are comfortable, and use it."

I felt I was losing support, so I added, "You know what? We will use it on a hide near the thumbnail. Thumbnail was a location in Ramadi that enabled us to shoot insurgents at up to 800 meters, extending our reach in the form of a BTHP 175-grain bullet. The hide was tactically advantageous and defensible. In addition, we had OP Hotel to fall back on in case they attempted to overrun our position.

Shadow 6 and his team were reaching the tail end of their deployment. Over the past few months, I was taking liberties to ensure our position here at Camp Corregidor remained strong.

A couple of days later, Bennett and I crossed the hallway to introduce ourselves to the Navy's EOD (Explosive Ordnance Disposal) sailors. The air in their hooch was thick with the distinct scent of tools of war and metallic instruments, the kind of place where innovation was born out of necessity. We laid out our intentions with transparency and purpose, explaining what we hoped to accomplish.

As we spoke, I noticed one of the sailors in uniform, his arms crossed over his chest, nodding with quiet approval. The demeanor of the group told me everything—they valued initiative, and our proactive approach had earned their respect.

What amazed me most was not their approval but the sheer simplicity of the device they introduced us to. After extensive training on explosives and complex procedures, I had expected a labyrinth of wires and circuits, something that screamed sophistication. Instead, the brilliance of its design lay in its elegance, a device so straightforward it was almost poetic in its functionality. Watching them demonstrate, it dawned on me how effective simplicity can be when paired with expertise.

> "It dawned on me how effective simplicity can be when paired with expertise."

As always, Bennett, my war horse, could not resist an opportunity to tinker. While I absorbed the finer details of the device's construction and initiation, he lingered behind, his eyes lighting up at the sight of new tools and gadgets. Typical Bennett—always drawn to "toys" that most would consider anything but. That was his charm. Where others saw danger or chaos, he saw potential, a puzzle waiting to be solved. Moments like this reminded me how vital personalities like his were to our mission—curiosity paired with ingenuity was a weapon in its own right.

After developing an idea and following EOD's instructions, a few days later, I found myself cradling the explosive device in both hands. It felt extraordinarily light. My heart pounded as I examined its simple design. It was larger than I had imagined when we had first pieced it together—a beast of a contraption that hummed with latent potential.

Like a gut punch, a thought hit me. *If I get shot in the back, will it explode? What's going to be left of me?*

My mind raced, playing out scenarios of what might happen if the worst came to pass. Images flashed before me—a sudden burst of heat, a deafening boom, and… nothing.

My next thought was even worse. *Could I ask one of my Soldiers to carry this?* The very idea made my stomach turn. They were my Soldiers, my responsibility. Asking them to shoulder this risk before I fully understood what it entailed felt like a betrayal of the trust we had built together. It felt cowardly. For a fleeting moment, I felt ashamed that I had even considered delegating this task to someone else.

I took a deep breath, inhaling the smell of our new chow hall and the stench of Ramadi. That grounded me. Resolve set in like steel. *If anyone is going to take this chance, it's going to be me.* The decision was terrifying, but this was my call and my burden. I adjusted the device carefully in my hands, feeling its awkward shape dig into my palms. The sensation emphasized the gravity of the choice I was making. Some people hate heights. For me, it was explosives.

As I secured the device to my gear, a strange calm settled over me—the essence of leadership—stepping into danger because it was the only fair choice. I didn't know if this gadget was worth the risk, but I knew one thing with absolute certainty—if I carried it and something went wrong, I would be dead. Asking my team to take that chance without fully understanding it was a gamble I couldn't take.

When the day of the mission arrived, we had planned and coordinated with all elements and selected our hide position. From prior experience, we knew a Sunni family occupied the home with two children, a boy and a girl.

From Shadow 2, SGT LL, and PFC PF would accompany us as security. As we stepped onto the road to leave the camp, the weight of the device pressing against my back was a constant reminder of the risk I had chosen to bear. I could feel it nearly sticking out of my ruck, awkward and conspicuous. *This thing is going to blow me to pieces before I get the chance to use it.* The idea gnawed at me, but I forced it down, steadying my nerves. I had to trust my preparation despite the cold edge of doubt.

I was confident in my idea, convinced it had real potential to change the way we operated. Glancing at the faces of my team, I could see skepticism. They did not say it outright, but their body language spoke volumes. No one truly believed it would work—or, if it did, they doubted its worth. The device was unproven, risky, and the consequences of failure unthinkable.

The burden fell to me. I wasn't about to let someone else pay the price for my conviction.

A quote from the countless military books I had devoured over the years played on a loop in my brain: "Audacity, audacity, audacity." I felt them like a chant with each step I took. Boldness is the spark that transforms ideas into success.

> "Boldness is the spark that transforms
> ideas into success."

Audacity—ruthless, relentless, and unshakable—was the only way forward.

We approached our position with stealth, confidence, and vigilance as we had done dozens of times before. Upon our arrival, we entered and cleared the home. We came across the woman of the house who had scorned Bush at our previous meeting. She appeared annoyed by our desire to once more use her home for hunting. She had a strong command of the English language. "Mister, why are you here? Go somewhere else!"

Her children woke up and immediately surrounded us—eight Soldiers in full battle gear. Her husband, with his hooded bloodshot eyes, merely gave us the once over. She was in charge.

In addition, an MAM (military-aged male) stood in a house. We had never seen him before, and he appeared agitated. This concerned me. I asked who he was.

The woman informed me that he was a family member but didn't give details about how he was related or why he was there. Suspicious, I searched him for anything that could harm the team. We followed SOP and cleared the house. Once we determined the house was secure, I remained on the first floor. I contacted the BN TOC to inform them of our position and our intent to continue the mission.

I extended my hand to her young son, maybe eight years old, shaking his hand gently, before turning my attention to Aiya who might have been twelve. Her warm, easy smile reminded me of my own daughter. She had a natural grace in how she carried herself. Her bright eyes and curious demeanor spoke of a spirit untouched by the harshness of our surroundings.

Lowering myself to one knee to meet her at eye level, I placed my hand over my heart and greeted her softly, "*Asalaam Alekum.*"

Radiating innocence and warmth, she replied, "Hello, how are you?" her English deliberate and careful, as though she were a student practicing for her teacher. Her effort caught me off guard in the best way, and I found myself smiling, genuinely moved. In that moment, she won my heart completely. In such a desolate place, her presence felt like a rare and precious gift—worth safeguarding at all costs.

One of the men approached. "SGT Martinez, she put up a new wall."

I looked at the lady. "We told you not to build the wall. You can't afford the materials, and we need you to cooperate with us, not work against us."

The brave woman stood her ground, her eyes unwavering, staunch as she faced eight heavily armed American Soldiers clad in full battle rattle. The fire in her voice cut through the tense air, defiant and unafraid.

"Ali Baba comes here and finds us helping you, my family dies. How can you help us?" she demanded, her words sharp with anger and despair. Ali Baba was their term for insurgents. "You want to help? Leave my house!"

Her tone carried the unbridled defiance of someone who had endured too much and refused to cower before overwhelming force. I gave her credit. She was determined.

I ignored her. Life had taught me not to argue with a strong-willed woman. You will lose.

"All right, guys. Set security. Have someone stay with the radio, and I will go upstairs with Bennett and set the device."

Ignoring their smiles, I reached into my ruck and pulled out the device. Security escorted the family to a room. The lady protested, but I shut out her words. I went upstairs and set the device. I placed it against the wall, following all instructions.

I ordered SGT LL downstairs. No one would remain upstairs until it detonated. I went downstairs and told everyone to get out of the common area. The husband was in the corner looking helpless, defeated, and depressed. He was no threat.

The lady again approached me. "What are you doing?"

I wanted to tell them to leave the house. Instead, told her, "I'm going to blow a hole in your wall large enough that we can move back and forth without any trouble. It's going to make a big noise. I want you to keep an eye on your children. Do you understand?"

She was clearly agitated. "Why, Mister? Why blow up my house?"

Why? Because she rebuilt the wall when I told her not to.

"Do as I say. Get your children and keep them away from the hallway." With that, I went upstairs and initiated the detonator. We had two minutes before it exploded. I wanted to run downstairs and hide. Instead, I marked the time and walked downstairs. SPC RT had set up a camera near the doorway to capture the explosion. Checking on the family, I noticed the lady and her children were not in the room, only the unknown MAM and the father.

"Where are Aiya, her brother, and their mother?" I demanded, concerned about their safety.

PFC PP replied, "They went into the kid's room. She insisted."

Knowing I had less than thirty seconds before the explosion and not knowing the effects, I searched the house. I found them huddled together on a bed. In less than fifteen seconds, I grabbed the two children, placed them on the floor, and covered their bodies with my own.

The explosion rocked the house. Immediately, we were enveloped in smoke. I heard stuff clattering to the ground, breaking glass, followed by laughter from some of my guys. I stood up, grabbed the two children, and quickly checked for any harm.

The lady was still sitting up in bed. She appeared to be in shock. I visually checked her, and she seemed uninjured. Stepping into the hallway, I saw the explosion's effect for the first time. As I ascended the stairs, surrounded by smoke, I found the wall and discovered a perfect 3x3-foot square cut into the wall.

Looking around, I was surprised at how much we had destroyed inside. Doors had blown off the hinges. Iron grates on the windows were lying in the courtyard, and the curtains were on fire. I headed downstairs and immediately encountered the unknown MAM.'

He said, "Why, Mister? Why?"

I retorted, "Shut your fucking mouth!"

He was wise and went back to the male homeowner.

In the kitchen, everything was in pieces: the cabinets, the table, and the chairs. The refrigerator door hung off its hinges.

I was suddenly aware of Aiya kneeling on the floor, picking up shattered dishware. Concerned she may cut herself, I grabbed her hand and quickly said, "No."

She was startled. "Aiya, I'm so sorry. Do you understand? Please, I didn't mean this to happen. Stay here. I will be right back." I motioned with my hands to remain in place.

When I returned to the kitchen, I opened my ruck like a desperate man and searched for the treats I always carried. I pulled out a gallon-sized bag of American candy. Aiya's eyes lit up. "Share this with your brother, sell it if you want to, but this is yours."

As I returned to the main hallway, I spoke to SGT LL, and we again surveyed the damage.

SGT LL said, "Well, if we don't have a fight on our hands now, we will soon. I think we can defend this."

Up to this point, I never felt confident in SGT LL. He was knowledgeable and gifted with the radio. However, his warrior spirit lacked common sense and tactics.

I told him, "Our perimeter is compromised. Let's go downstairs. I will explain to the female owner and leave her a note to go to the Government Center to collect something for the damage. Ensure everyone has accounted for sensitive items, and no one is hurt."

When she saw us, the woman's voice became shrill and loud. Trembling, she pleaded with us, her cries filling her shattered home. Her hands gestured frantically toward the garden outside, where her vehicle sat damaged, covered in debris. The weight of her distress pressed heavily on me. Each word exposing the anguish of someone caught in the collateral chaos of war.

Her cries cut through my composure. Hoping to offer some semblance of restitution, I scribbled a note, careful not to reveal too much about who we were, and wrote that the damage had been accidental. I requested that the Government Center address her losses. The words felt hollow even as I wrote them.

When I handed her the note, she took it with shaking hands and read it slowly. Already etched with grief, her face crumpled further. Instead of offering comfort, the note deepened her despair—an unadulterated confirmation that her cries for help had yielded little more than empty words scribbled on paper. Her anguish stayed with me, a haunting reminder of the human cost of every action, no matter how unintended.

Her face twisted and red, she screamed, "The government center? They will kill me. They will see me and kill me. Please do not leave. We have lost everything! Why did you do this? WHY?"

As our team gathered near the door, I felt compelled to explain that the extensive damage from the blast had been unintentional. But my words felt hollow before they left my mouth.

In full-on hysteria, her cries cut through the air like a blade.

Aiya clung to her side, a picture of quiet strength amidst the chaos. She wasn't afraid. At least, she didn't show it. Soft and soothing, her voice carried maturity far beyond her years as she tried to comfort her mother. Yet, her efforts did little to stem the torrent of grief and fury that consumed the woman. The sight of Aiya's small hands on her mother's trembling shoulders, her calming words swallowed by sobs, was almost too much to bear.

The moment hung heavy with unspoken truths, the weight of unintended destruction, and the unrelenting reality that spared no one—not even the innocent. Aiya's quiet bravery underscored the heartbreak she was forced to bear.

We were moving through the courtyard, the team filing out through the gate one by one with the woman following us. I was the last man, trailing behind the others, when Aiya's mother suddenly grabbed my arm. Her iron grip and her desperation stopped me in my tracks. She was begging me for something I could not give her.

Trying to shake her off, I said, "Lady, let me go. I have to go!"

She dropped to her knees, her hands clutching my arm like a lifeline. Tears streamed down her face as she sobbed, her pleas growing louder, more frantic. Her despair caused me to doubt why we were here in Iraq. The surprise of her actions left me frozen for a moment, torn between the weight of her anguish and the reality of the situation. Hajji was coming, and we had to get out.

The scene felt like a crossroads—a split-second decision with no clear path. Her cries echoed in my ears. I could feel the team moving farther away, their steps fading into the distance, while her grief rooted me in place, forcing me to grapple with the impossible. How could I walk away from someone so utterly broken?

PFC PF's strained voice brought me around. "They're getting closer. It sounds like a big crowd."

I could no longer concern myself with Aiya's mother. Her actions were jeopardizing the lives of my team.

Feeling regret, I struck Aiya's mother in the face with my open hand. The strike landed hard, and she fell on her back. I looked at Aiya's shocked face.

"I'm sorry," I whispered.

I moved quickly to the outside gate where PF still stared at the lady on the ground. In an assertive tone, I said, "Let's go."

He didn't move. I said, "LET'S GO!"

We hurried across the street and returned to OP Hotel. I never saw Aiya again. I had failed my mission and shattered perhaps the last remnants of humanity within me.

Debriefing, we concluded that our device had lethal potential. It was raw and needed refinement. Where others might see imperfection, we saw opportunity. Partnering with the Navy's EOD, we made the necessary adjustments, tweaking its design until it became more precise, giving us another instrument to hunt and kill AQI forces.

The plan was cold and deliberate. We would integrate this deadly tool and use their own tactics against them to send a message in blood that would destroy their will to fight and bring on a sense of doom. The goal was simple and grim—an increased enemy body count. In war, necessity and brutality are two sides of the same coin.

Kill the Dump Truck Driver

OP Hotel became our home away from home, a fortress in the heart of a broken city. Our mission itself settled into a dangerous rhythm that felt almost routine. From here, we stacked bodies, a grim testament to relentless combat. All the sniper teams, including SSG James Gilliand's Shadow 6, were actively defending the building against an enemy intent on its destruction.

The wear and tear of war was impossible to ignore. The once-sturdy structure bore the scars of countless engagements. Walls riddled with bullet holes told the story of endless small-arms fire. Debris covered the yard from rocket attacks and garbage blown from the city streets. Its walls bore scorch marks from enemy weapons, leaving jagged cracks in the wall. Some rooms were utterly unusable. Their ceilings were exposed rebar, fractured from repeated blasts. Stairs were cracked and littered with wreckage. The building was holding on by sheer willpower, like we were.

Somehow, against the odds, OP Hotel stood. It remained a defiant monument to our resilience. How much more could it take? How much longer before the next hit would bring it all crashing down?

An Iraqi Shia platoon occupied the floor below us, though to call it occupied felt generous. I never heard from them, never spoke to them. Every time we caught a glimpse inside their space, they clearly operated in a world far removed from ours. By grunt standards—a low bar, mind you—their area was a *bona fide* disaster zone. Trash littered the floor.

Gear was helter-skelter. The smell was enough to turn your stomach. Lack of discipline radiated from every corner of their quarters.

To me, they were little more than a token presence, a speed bump for the enemy to crush on their way to us. Cannon fodder. That harsh assessment was shared by many who watched them flounder in the basics of soldiering.

Later that year, we began the slow process of handing over more responsibility to the Iraqi Army. It was a rough and uneven transition, filled with setbacks and frustrations. Progress came inch by inch, each step a test of patience and persistence. In those early days, I saw only chaos and ineptitude.

During the Global War on Terror (GWOT) between 2005 and 2006, the Rules of Engagement (ROE) defined the fine line between life and death, guiding how and when force could be used in combat. These rules were meant to balance operational effectiveness with the ethical and legal obligations of war. In our Area of Operations (AO), the ROE evolved constantly, often in response to the enemy's shifting tactics, techniques, and procedures (TTPs). Adapting to these changes became a critical part of survival and success. While this might sound chaotic, it was a grim necessity. In a battlefield where the enemy blended seamlessly with civilians, the ROE served as both a guide and a weapon.

For example, one particularly strict set of ROE emerged to counter specific threats we faced daily. Motorcycles with two riders became instant targets if the passenger sat backward, a common indicator of a spotter for an ambush or an IED detonation. Our orders: kill them.

Military-aged males (MAMs) standing on rooftops observing American patrols were presumed to be enemy scouts: kill them. Anyone observing OP Hotel was treated as a direct threat: kill them. A black mask—a symbol of insurgent groups—marked the wearer as hostile, no questions asked: kill them.

MAMs who peeked around a corner more than three times, a behavior suggesting an ambush setup, were considered enemy combatants: kill them. Using a cell phone near an American patrol, a potential for detonating IEDs or coordinating attacks: kill them.

These rules often had shockingly short lifespans, sometimes lasting less than twenty-four hours. They were reactive measures tailored to counter immediate threats, leaving no time for clarity or debate. This fluidity demanded absolute focus and adaptability from every Soldier. Yesterday's rules could be irrelevant—or even dangerous—by morning.

The shifting ROE reflected the unique challenges of asymmetric warfare, where the enemy operated without rules, blending into the civilian population and exploiting every advantage. For us, these rules were survival strategies, brutal but necessary in a war where hesitation could mean death. They also served as a grim reminder of the moral weight we carried, making split-second decisions under shifting guidelines. Who could tell what was right and what was wrong?

Of all the amended Rules of Engagement (ROEs) I heard and acted upon, three stood out as stark and immediate: black masks, cell phones, and an individual digging a hole near a path or roadway. Each carried a distinct threat, and each demanded decisive action.

On this particular day, we were stationed at OP Hotel, where the enemy preferred to strike in the light of day. Anticipating this, I allowed my men to rest during the night, knowing that by morning, we would be fully alert and ready to deliver our own lethal response.

As the first light broke, I headed downstairs to the TOC (Tactical Operations Center) to get the latest intelligence from the NCOIC. There, I was briefed on an ominous new development: S2 had intercepted enemy communications reporting that several stolen dump trucks had been packed with explosives, their purpose chillingly clear—level OP Hotel in a single devastating strike.

The threat was immediate, and the response was swift. A new ROE crackled over the radios to all CF operating in Eastern Ramadi: any dump truck seen within the vicinity of OP Hotel was to be destroyed on sight, no hesitation, no margin for error.

To bolster our defense, Battalion TOC deployed tanks, positioning them about 800 meters west of OP Hotel near the circle, in close proximity to the Government Center. Held and fiercely defended by the Marines, this stronghold would serve as a critical buffer against the looming threat. The presence of the tanks was a show of force, a deterrent to any enemy who might think twice before approaching with one of their explosive-laden vehicles.

Every sound, every movement outside the perimeter felt magnified, each potentially heralding the arrival of a weapon to annihilate us. In that moment, the weight of the ROE turned cold directives into life-or-death decisions.

SPC BY and SPC Bennett were assigned to cover our western approach. SPC RT held watch to the south, while I kept my focus on the northern perimeter. I stepped downstairs briefly. As I returned, the loudest explosion I had ever heard in my life shook the building. I grabbed the doorjamb, steadying myself, my heart pounding.

"SGT Martinez, we're okay—it's one of ours," SPC RT said, his voice calm, even managing a slight smile.

"What the hell was that?" I demanded, still shaken.

"It's a main tank gun," Bennett chimed in eagerly, practically buzzing with excitement. "It just hit a dump truck that was speeding east inside the circle, heading toward our position." The awe in his tone made it clear he was impressed with the Abrams A1 tank's devastating power.

Snapping into focus, I turned my attention to the radio. The static-filled chatter barely registered as the tank's coaxial gun barked out a stream of hate. Moving quickly to our viewing loophole, I peered out and spotted another dump truck attempting to maneuver past the tank. The vehicle swerved erratically as a line of coaxial rounds chased it down.

The truck's evasive maneuvering pushed it dangerously close to our position. Without thinking, I took control of the situation, my voice firm and decisive. "BY and Bennett, kill the driver!" The words left my mouth with urgency and authority. Repeating myself for clarity and focus, I declared, "I have control, I have control, I have control."

As the truck roared around the bend, speeding from west to east, my voice steadied into a calm, commanding tone. "Execute, execute, execute."

Almost immediately, two shots rang out. Time slowed as I watched the dump truck grind to a halt, coming to a complete stop seconds later. The radio crackled again, this time with the tank commander's voice filled with surprise.

"Shadow has killed the driver and passenger—no further damage."

Through the loophole, I stared at the truck, my eyes fixed on the windshield. Two perfect entry points were clearly visible. Lowering my binos, I grabbed the spotting scope for a better look. The driver was slumped over, partially leaning toward the passenger seat, though the passenger was no longer visible.

Keeping my composure, I retrieved my camera and snapped a photo of the scene. We had just killed a driver and passenger—deemed threats—traveling at over forty miles per hour, at a distance of 730 meters. Each sniper had fired only one shot, and both had hit their mark.

Heading downstairs, I relayed the details to the NCOIC, who passed the report up to the BN TOC. The amended ROE regarding dump trucks proved its worth that day, though by the end of the operation, it was eventually rescinded.

We had no further incidents that day, but the image of that truck and those two shots stayed with me—a vivid reminder of the precision, decisiveness, and gravity of the choices we made in combat.

The Fight at the Stadium

Death Dealer Company 2/69 was determined to leave Iraq with a statement—a final, resounding strike that would echo through the ruins of Ramadi. Fueled by this motivation, they devised an audacious plan to seize the stadium in Mula'ab and launch a relentless series of combat patrols for three grueling days. The stakes were high, and the mission had a clear objective: take the fight into the heart of enemy territory, a neighborhood so notorious it earned the chilling nickname The Badlands on the map in the war room of Full Metal Jacket.

The orders came down from SSG Gilliland, his tone unwavering as he summoned us to FMJ. The plan was bold, almost reckless in its simplicity. We would insert with BFVs and Abrams tanks, storm the stadium, and hold it as our stronghold. From there, the assault would expand into Mula'ab. Snipers, the scalpel of urban warfare, were tasked with filtering into the labyrinth of alleyways and crumbling buildings, setting up hides to systematically eliminate hostile targets and cripple enemy resistance.

This was no routine mission. It demanded flawless coordination from all four teams, a unified effort to maintain relentless pressure in a neighborhood that thrived on violent chaos. Mula'ab was a maze of hostility where every shadow held a threat, every corner a potential ambush. Tension crackled through the air as we prepared, a quiet intensity that sharpened our senses. We stripped and reassembled our weapons with meticulous care, distributed ammo with precision, and examined every piece of gear as if our lives depended on it—because they did.

The clock was merciless. Zero hour was set for 0330. The countdown bore in on us. Each passing minute stretched long and heavy, filled with unspoken thoughts of what lay ahead. This was the kind of mission that left its mark on the body and the soul. Every decision, every movement could affect the survival of the entire team. We were descending into a crucible, where failure was not an option.

We adhered to the plan, executing each Standard Operating Procedure (SOP) with precision as we gathered at the COP. The BFVs stood in neat formation, lined up in order of march (OOM). Their engines hummed faintly in the cool morning air, the armored power that would lead our push into enemy territory. Crews moved with purpose, some making final checks while others stood in quiet clusters, their low conversations blending into the early dawn.

My rucksack weighed heavily on my back, a symbol of the grueling hours ahead. In addition to the standard gear, I had packed snivel gear for cold weather and a sleeping bag. The weather was changing, and the early morning chill prompted many of us to pull on our black fleece jackets. As I stood there, I couldn't help but think how absurd it was that we wore black at all. As tactical Soldiers, we knew black was unnatural, a color that stood out even in the shadows. Green or brown would have been better choices, but someone somewhere had cashed in on polar fleece contracts, and we bore the brunt of that decision.

The hum of conversations floated around me, snippets of plans and reassurances, but I kept my thoughts to myself. I didn't want to be that Soldier—the one always asking, always probing. I knew my place. I was surrounded by professionals, men who carried themselves with the quiet confidence of experience. For me, this was a time to learn, to watch, and absorb.

While waiting for the green light, we heard the E4 mafia adding their twist and narrative to any mission. The legendary "E4 Mafia" was less an official rank and more a state of mind. Every unit seemed to have them: seasoned specialists who somehow know exactly how everything works, and more importantly, how to avoid everything that doesn't need to get done right now. They're masters of timing, vanishing just before a detail gets handed out and reappearing like nothing happened once the dust settles. The myth says they run the military from the shadows—not with authority, but with experience, sarcasm, and an unmatched ability to "sham" with precision. While officers make the plans and NCOs enforce them, the E4 Mafia quietly keeping things from completely falling apart… or at least making sure everyone gets a good laugh along the way. That morning was no different.

"You think they will call this off?" asked a voice in the dark.

"You know Death Dealer, 'CONTACT – BREAK CONTACT,'" followed by the Soldier's laughter.

Another voice replied, "I could give a fuck, man. This is Dealers' show. Fuck them! We are not making a difference out here."

Although one remark appeared to be true, labeling the entire Death Dealer company in such low terms wouldn't be fair. They had a great corps of NCOs and officers. The repetitiveness of fighting on the same streets was wearing thin on those who had to leave the wire. It left me with the impression that the general attitude was "we just want to do our time and get home."

As we continued to wait in the dark with the vehicles running, I heard reports of the clearance route suffering casualties. This incident caused a stir of doubt.

"Well, they are certainly going to cancel this…" another one added.

We were supposed to insert under the cover of darkness, but Murphy's Law entered the game. Once the first shot rings, Murphy will show his face.

The order came through: mount up. We knew full well the risks ahead. The route clearance teams had already taken casualties. The odds of us getting hit were high. Every man understood this wasn't just another ride. It was a roll of the dice through a kill zone. I climbed into the BFV, squeezing into the tight space built for Soldiers but not for comfort. The compartment was stifling, barely large enough to fit us all, our gear jammed against one another like puzzle pieces forced into place. The only light came from a small portal on the ramp, casting an eerie glow that did little to illuminate anything.

The engines roared to life, a metallic growl that vibrated through the steel hull. The radio crackled constantly, the net flooded with transmissions as commands and updates bounced between vehicles.

With a jolt, the BFV lurched forward. The movement was abrupt—short stops and sudden surges that sent us rocking against each other in the darkness. Soon, the vehicle settled into a steady rumble, carrying us toward Mula'ab and whatever lay ahead.

I prayed, my thoughts a jumbled mix of desperation and resolve.

Oh, Christ, please let me get there in one piece.

I just want to get the fuck out of here!

Why didn't we insert on foot?

Words tumbled through my mind on a loop of fear and frustration. At moments, when the jolts of the ride or the ominous radio chatter caught me off guard, I muttered, "Jesus H. Christ!" It was half-prayer, half-curse, but it felt necessary—an outlet for the storm raging within me.

Dwelling on it served no purpose. We were committed. This fight was inevitable. I forced myself to focus. The plan was in motion, and my doubts didn't matter. This was not about liking the mission. It was about winning the fight. The only way to do that was kill as many of the enemy as I could.

I did not mind the fight. Hell, part of me welcomed it. But the way we were thrown into the mix sat wrong with me. I knew better than to voice it. After all, we were National Guard, and these guys were Regular Army. That distinction lingered over us, a constant reminder to stay in our lane.

I tightened my grip on my weapon, swallowed my frustration, and braced for the storm. Whatever waited for us in Mula'ab, it wasn't going to wait long.

When the ramp dropped, the first thing I saw was a decaying wall, its crumbling surface showing the harsh realities of this war-torn city. As always, stepping out of the BFV was a commotion of gear snagging, bodies jostling, and everyone trying to get clear in the confined space while keeping their weapons ready.

The dismount left me on edge. We were exposed the moment we left the relative safety of the vehicle. As I struggled to secure my gear and scan the surroundings, an uneasy realization settled in—this operation might be a bad idea. Supposedly a prominent structure for the city of Ar Ramadi, the stadium looked like a dilapidated high-school ballfield, flanked on three sides by multi-story buildings. Those buildings loomed over us, ominous and menacing, offering perfect overwatch for anyone with hostile intentions. It was a tactician's nightmare.

CHAPTER 7

CAPTAIN RARD AND ANGLICO

The Stadium inside Mula'ab Ramadi 2005

Task Force Dark Eagle (Marines and Soldiers) inside Mula'ab

We moved in Ranger file. That means moving in a straight line, one person behind another, usually through tight or dangerous terrain like woods, alleys, or narrow paths. Each person follows directly in the footsteps of the one in front of them. This helps keep the group organized, reduces noise, and—most importantly—limits exposure to threats like ambushes or explosives. We were trying to link up with the rest of Shadow. The environment made coordination nearly impossible. When it became clear the link-up wasn't happening, we pushed into the stadium, seeking high ground. Gilliland quickly assigned teams and sent us to our respective positions. To my north were SGT FN and Bennett.

Southwest held SHADOW 6, 2, and 3. The Iraqi Army (IA) was stationed to the southeast. At the heart of it all, under the press box, was the Tactical Operations Center (TOC), the nerve center of this operation.

On the soccer field below, vehicles were arranged left to right in a neat, almost ceremonial row. It was unnerving. The scene reminded me of archival photos of Pearl Harbor, where rows of ships sat vulnerable, waiting for the storm to hit. *Too easy*, I thought, *an invitation for disaster.*

Directly ahead at my 12 o'clock, a pair of tanks sat behind a stadium wall, their guns trained south. To my immediate right was a line of M240s and M249s, manned by Soldiers from the Pennsylvania ARNG Animal Company. To my left, a group of Marines maintained their post. The road in front of the tanks—a straight, open stretch extended over 300 meters. It looked like a runway, a perfect approach for a VBIED. I felt a chill go through me.

As I surveyed our position, a frustrating oversight became painfully clear. The walls around the stadium were solid, providing decent cover, but the lack of loopholes forced us to expose ourselves to observe. Every time I poked my head above the wall, I felt like a target. It was a rookie mistake, and I knew it. We hadn't brought the tools to create proper observation points, and now we were paying for it before the fighting began in earnest by exposing ourselves to incoming SAF or worse.

Still, there was no time to dwell on regret. Mistake or not, this was the fight we were in, and it was our job to win. Tension filled the air, the silence broken only by the occasional crackle of the radio or the shuffle of boots on concrete. Anxiety churned in my gut, but I pushed it down. Mistakes might haunt you later, but in the moment, survival demanded focus—and I was determined to make it through.

"This was the fight we were in, and
it was our job to win."

Day 1

Gilliland killed two insurgents attempting to man a room with a weapon.

Numerous motors struck to our north and east injuring 6 IA's.

Bennett killed an insurgent with a spectacular head shot at 100 meters.

A Huey landed and carried out the six wounded.

One of their mortar rounds failed to detonate, and it was discovered by Soldiers near the TOC. Had it exploded it would have put a serious hurt on our leadership. Later, the mortar was intentionally detonated (control detonation) by Navy EOD.

I observed a maroon-colored van with black windows enter the roadway to my 12 o'clock. Its rear end was angled toward us, and I observed the rear passenger side vehicle window pop open. I thought this had to be an insurgent trying to snipe someone. When the window opened, I could see sunlight from the front windshield entering the interior of the vehicle. I went into a prone position and sighted my weapon on the window.

The back of the van was moving up and down as if someone was jostling around inside. No way this was two teenagers caught in a moment of lust who failed to notice as an American Infantry unit with armored vehicles stormed into the stadium. Furthermore, this wasn't America. This was Ramadi.

When the person entered the sunlight, I took my shot. My weapon bounced, and it took me a little longer than usual to reacquire my position. I looked at SPC JY. He had not bothered to grab the spotting scope. As soon as my shot went out, we heard another shot. The radio came alive with traffic, and someone wanted a report. I reported my situation and heard nothing further about it. Gilliland came over the net and told me to relax and wait until we have a clear and definitive situation. I should avoid jumping to conclusions.

I looked at SPC JY and commented, "I only took one shot. Why does he think I'm jumping to conclusions? Someone shot at us about the same time I took my shot."

In his typical manner, SPC JY shrugged and said, "I don't know."

I decided to locate Gilliland and clarify myself. When I found him, he was located near the catwalk to my west. I did not want to come off angry, so I made an extra effort to act professional and make my inquiry. Apparently, Gilliland heard two quick shots and thought I had taken both of them. When I informed him, I had only taken one, Gilliand smiled. "No worries, it sounds like someone out there was trying to get some. I guess you beat him to the punch line. Did you kill him?"

"I don't know, the long gun jumped on me when I broke the shot." Feeling confident, I returned to my position and found SPC JY searching the area (Good man).

When darkness arrived, I moved off the wall and located a place to sleep inside the press box.

Early the next morning, SPC JY and I discovered a folding couch and carried it up to our position. We set it up and began to scan.

Day 2

I encountered Top (a term reserved for First Sergeants, adopted from the German Army), my 1SG, Richard Hughes from the NJ ARNG. A former Marine and nearly fifty-eight years old, he was fighting with the Engineers.

"Top, what the fuck are you doing here?" I said, surprise in my voice.

"I couldn't stand the admin bullshit at Camp Ramadi, so I found a way over here on the east side and attached myself to an engineer unit assisting you guys." He was sweating profusely and looked perfectly at home.

"Top, don't get yourself killed." I was genuinely concerned for his safety.

He put his helmet back on and said, "I'm going out there again," and headed back to ground level. Crazy fucking Marine.

I decided to locate Gilliland and asked him for some clarification on our mission. I felt we were nothing more than another set of grunts with special guns. I didn't mind it, but I wanted to get out into the city and hunt insurgents down as they tried to move on us here in the stadium.

In my search, I found two Shadow teams inside the hollowed-out portion of the staircase beneath stadium seats. Several Shadow team members were still sleeping. Sharing their quarters was the carcass of an animal deep in decomposition. In the end, strip us of our badges and weapons, and we are no more than simple infantrymen. As snipers, we just bring a little spice to the death ground.

Gilliland leaned in close and shared his assessment with me. "Look, I think the DD6 (Death Dealer Company Commander) took on more than he can chew. He intends to remain inside. We are now ordered to stand fast and assist in perimeter security."

I couldn't hide my disappointment, "What the fuck? This is nothing more than a slugfest, and all we are doing is sitting in a vulnerable unsecured position. We have buildings ringing us on three sides."

"Hey, man, no worries. Keep low and don't take any unnecessary risks." Gilliland gave me that smile that said it all.

I didn't want to sound like a whiner, so I said, "Roger that, brother." I looked around again and returned to my position.

SPC JY met me with questions, and I gave him answers, ending with, "We are no longer snipers, just grunts manning a wall. This harassment is going to get a whole lot worse before we leave. Watch your back and make the best of it." I returned to the wall.

We began to take small arms fire (SAF) from a house at our 1 o'clock, in our southern direction. We gave an instant reply but to no avail. Two LTs from 2/69 walked over to hatch a plan together to take out the enemy machine gun. It was beautiful in its simplicity. From their position, they walked in two tanks and fired main gun rounds into the house. Their guns brought a cry of joy from the grunts and a loud silence from the smoking house.

A nineteen-year-old combat engineer crossing the street to return to the stadium was shot by an enemy sniper. The round hit him squarely in the chest. I observed him running inside the walls, assisted by other Soldiers. I decided to check on him, and I located him inside the perimeter.

His SAPI plates saved him. The round hit the top edge and turned right, entering underneath his blouse and coming out at the bicep area of the sleeve. When I spoke to him later, he said he was looking toward where he was running and never saw a flash. Afterward, I saw him going back out into the city. *Amazing* is the only word that rings inside my head—a testimony of the men and women fighting AQI.

Iraqi soldiers abandoned their position and began to play soccer on the field. We watched them in astonishment and shook our heads.

When darkness fell, once again SPC JY and I returned to the press box to sleep.

Day 3

I woke up a little stiff, annoyed, and hungry—the plight of a grunt. SPC JY and I returned to our position. As the day passed, we received machine gun fire from a rooftop at my 2 o'clock. Some of the rounds hit near us. Honestly, I couldn't help but laugh at the insurgents fucking with an Armor company reinforced!

We improved our position and knocked out oversized loopholes with a sledgehammer borrowed from the Engineers. Bennett visited us a couple of times, and I could tell there was some drama going on between the team members of Shadow 4. I figured they would work it out themselves. As long as it wasn't carried out on a mission, I wouldn't concern myself with it.

The constant machine gun harassment began to grind on my nerves. I debated whether to take initiative and spread heavier guns along our wall. I shared this thought with SPC JY. He shrugged and looked away, his typical sign for agreement. As I surveyed the area, I identified the areas within my battle space where we could be shot, and I realized I couldn't do this myself. *Fuck It!*

I went to Animal Company and said, "Hey, listen up. We have to spread some of you guys and the machine gunners out. This is what I want… Give me a team here 240 and 249." Pointing at a Soldier, I said,

"You, come down here. Hey, you move a little farther to your right, and you guys maintain."

When no one stopped me, I went on. I went to each Animal Soldier along the wall and pointed out danger areas—base of trees, corners of building, open windows, open doors, and dead space. The enemy machine gunner was still tapping out rounds in our direction, and this provided all the justification I needed.

Finally, I said, "All right, does everyone know where to shoot?" Heads swiveled in my direction. Everyone appeared calm and focused. "Great, once that fucker starts to shoot at us again, I want everyone to shoot a few rounds into their respective danger areas. Big guns keep the guns talking. No long burst." Drawing in a breath, I went on, "Keep this phrase in your heads as you pull the trigger, *Die, motherfucker, die.* It will help you with sustained fire."

While talking, I was standing. I knew the danger of exposing myself, but I figured this fucker hasn't been hitting anything but a wall, so fuck it. Suddenly, I saw a flash to my 2 o'clock front and immediately rounds hit the flag post at my 2 o'clock. One round hit the wall directly in front of me. That was all I needed.

I projected my voice. "ANIMAL, engage your targets!" All the guns opened up in small bursts. With such a heavy wall of lead before me, I stood up to shoot where I believed the machine gun was located. Standing up and exposing my front was a controlled and calculated risk.

What happened next was surprising. We started receiving SAF from those areas. As the insurgents returned fire, our volume of intensity increased. We had a full-fledged firefight. Both sides were determined to kill each other, and no one was yielding space.

As the intensity continued to rise, Animal performed flawlessly. Men were shouting, changing magazines, and covering each other. The 2 LTs from 2/69 (both warriors) arrived, so I backed off the wall. Additional Soldiers ran toward our wall to get into the fight.

"SGT Martinez, whadya got?" LT VV asked. He was fair-skinned and tall with close-cropped hair. He carried himself with an air of confidence that told people he was going somewhere. I had tremendous respect for him and his intellect… and his fighting spirit.

I low crouched down the stairs until my head was below wall height, so I could stand up and speak. "Sir, we started to receive heavy machine gun fire from our 2 o'clock position. In order to kill them, we had to secure the front. I had Animal's heavy guns lay down cover, so I could lean out and kill them."

LT VV smiled to himself and rubbed his chin. "Did you kill him?"

I responded, "I don't know, sir, but from the sound of it, I think we caught them moving on us. It could have been a ruse, possibly keeping our attention with the machine gun while they moved on our front. I think we stole their initiative."

As we were speaking, I turned around and moved up a few steps and caught sight of movement slightly to my 11 o'clock. A man came into the courtyard. He tried to run towards his vehicle, but he was turned back by heavy fire. I was surprised to see him come out again.

This time I climbed higher on the steps which gave me a bird's-eye view into his courtyard. He was carrying a rifle at port arms. Without a word, I leaned against a cement pillar and placed the red dot of my combat close optic in front of the driver's side door.

As soon as his head entered the optic and touched the red dot, I squeezed the trigger. He stopped, stood in his tracks, and his arms fell. Dropping the weapon, he crumpled forward. I brought my rifle down and scanned for other targets. Everyone on the wall was shooting. I returned to the LT and reported my actions.

Slowly, the rate of fire decreased. Finally, silence hung in the afternoon sun. I stared down the street at a donkey limping near a corner. A man with a long white beard was dressed in white with a *shemagh* that was white with a black headband. He carried something in his right hand and waved it above his head, shouting something.

Initially, I thought he was doing a one-man battle damage assessment. Then, I realized he was trying to inspire another fight from his brethren. His message was for his brothers to kill us. As long as he breathed, his voice was as lethal as a bullet.

Before I could give any commands, he walked away. I called out, "Hey, everyone. See that old man?" Some heads nodded. "That's a religious leader trying to inspire his followers to fight and kill us. If he shows his fucking self again, kill him!"

As if on cue, he reappeared. Again, I was mesmerized by his courage. I didn't shoot, but I heard rounds screaming down range. He was literally walking from right to left. For a moment, my imagination was getting to me. No rounds were hitting him. Was this an act of God?

He stepped up on the sidewalk, and I thought he was getting away. I brought my rifle up to my shoulder and leaned against the pillar for support. In a low voice, I talked myself through my fundamentals to stay focused. I placed the red dot just before the corner: the ambush technique. As he appeared inside the red dot, it was a textbook center-mass shot. I squeezed the trigger and saw the shot hit. He stumbled, and I briefly saw his man-dress ripple along the front of his body. A thin red line appeared near his hip and upper thigh as he fell behind the corner out of sight.

When I brought my weapon down, a voice loudly exclaimed, "I got him."

I looked at SSG BN who had a smile, full of pride. We made eye contact, and he repeated himself. I did not challenge him. I did not care who got the credit. This was not a game. My objective was to kill as many of the enemy as I could. If another Soldier felt it was his kill, so be it.

A few hours later, we broke our positions down and climbed aboard the BFVs. After a quick prayer, I found myself back on COP. It was over. We abandoned the stadium, and I couldn't find victory in the operation. My personal victory was that I had won another gun fight, and I was alive and breathing.

A few days later, a Marine (MTT) team informed me that they had spoken to an informant, and he had reported "thirty-nine bad men" were killed by our rifles.

Task Force Dark Eagle

December arrived. Christmas holidays were approaching. I, for one, could not have cared less. I missed my kids—three boys and a girl—but the holidays were just another day on the calendar. I was bored and decided to walk over to Shadow's quarters inside Full Metal Jacket.

As I approached in the darkness, a group of Soldiers stood in a semi-circle talking to one another. As I got closer, I recognized them. Gilliland appeared to be in deep thought pondering a response to a question.

Not wanting to miss anything, I asked, "What's going on, guys?"

SGT SH spoke but not to answer my question. "I don't know about them. I don't like going out with people whose patrol techniques may be poor. I've gone out with ANGLICO (AIR NAVAL GUNS LAISION COMPANY), but I don't feel like going out with them." SGT SH was one of the best in the section, and his point of view had validity. I also felt it wasn't fair to have these guys go out when they were within days of going home, and Shadow 4 was doing nothing with several months remaining in country.

Interested now, I asked, "Are you talking about Marine ANGLICO?"

Gilliland said, "Yeah, we have a mission with them."

I said, "Well, James, if you don't mind—and SH—I'm not trying to take your thunder, but I'll grab my guys, and we'll go."

SGT SH responded, "I don't care."

I said, "James, you cool with it?"

He nodded. "No worries. I'll introduce you to the ANGLICO team. They are housed here in FMJ."

"Great." SGT SH said and walked away.

As we walked over, Gilliland said, "Look, I'll introduce you. When we get back to our hooch, I'll explain the mission to you. The only coordination you'll have to do is submit a new trip ticket to the BN TOC and do final coordination with ANGLICO."

"Roger, no problem."

ANGLICO was located on the opposite side of Full Metal Jacket. We knocked on the door and entered. Seated inside were several young Marines in various stages of uniform.

Gilliand asked no one in particular. "Evening, gents, is CPT Garrick Rard here?"

A young Marine immediately stood up, "Yes, Staff SGT. I'll go get him."

A few moments later, he brought back a tall, blue-eyed, blond-haired Marine Captain who was built like a pro football player.

Introductions were made. I learned that CPT Rard used to be a Harrier pilot. The entire exchange was fairly casual. I explained that I had to leave to prep my team. Around 0001, I would return for final coordination.

CPT Rard was an imposing, soft-spoken giant with a natural authority. He approved the plan, and we began a fast three months as Task Force Dark Eagle. In that short time, TF Dark Eagle would be responsible for disrupting and systematically suppressing IED teams in the enemy's backyard. The bodies were going to drop. Either they would land in one piece by a sniper's round or in hundreds of pieces by the ANGLICO signature move—death from above. ANGLICO was comprised of two four-man teams, each led by a Marine CPT and a Marine NCO with two non-rates.

Gilliland and I returned to Shadow team's quarters where he provided the mission statement. On our first mission, our objective was located approximately 600 meters from the front gate of Corregidor. We knew the house. At the last minute, I added a sledgehammer to our mission-essential equipment.

At an agreed-upon time, Shadow 4 walked over to ANGLICO's hooch, and the teams were introduced to each other. I felt confident about working with ANGLICO. I had been in the Marine Corps. My experience with their unit was limited, but their professionalism and capabilities were well known. To me, they were another branch of Marine Recon.

CTP Rard said, "Okay, gents, here's the route and our order of march (OOM)." We discussed our method of entry into the courtyard based on what we may encounter. In addition, we designated entry teams and security teams. The CPT explained the two-man rule and ensured everyone understood. He provided QRF (Quick Reaction Force) and the frequencies. Combat loads were put out, and actions on the objective were articulated. We closed the final coordination with rules of engagement (ROE).

CPT. Rard added an element we never had before, a canopy of comfort. ANGLICO was able to put eyes in the sky and deliver real-time information as we moved. This tactic enabled us to move against any perceived threat before it could do us harm. It also enabled us to set up a hasty ambush. Ambush the ambushers.

That night, we moved into our position. For the next twenty-four hours, we sat and watched the Mula'ab neighborhood from a northern position looking in. This neighborhood was a breeding ground of insurgent activity, making the threat level remarkably high.

Leading to the house, we ran into one of those situations that are never addressed in training.

Marine Lance Corporal Jon Cureton was your typical hard charger, tough as nails with a no-nonsense approach, confidant and competent with a capacity for violence, a term of respect. I heard his voice in a hoarse whisper, "SGT Martinez!"

'What's up, bro?"

"The house next to us saw us enter the home."

"Are you sure someone saw us?"

"Yes, SGT. He walked out the front door and watched us hop over the wall."

"Shit. Okay, let's get inside, and I'll figure this out." Debating my options, I decided the best course was to discuss this with CPT Rard,

I informed him of our dilemma. We stood quietly for a moment and almost simultaneously began to speak.

"You have an idea, SGT?"

"Yes, sir. Our options are either abort or Charlie Mike (Continue the Mission). If we choose Charlie Mike, then I say let's put an entry team together and go next door and grab the family." I was expecting a big "No, Harry."

Instead, I heard, "All right. Two of mine and two of yours."

I said, "Roger, sir." The Marine Corps is never one to back away from a fight. True to form, ANGLICO would kill as many with their rifles as with their fast movers (Marine Corp fighter jets) in the weeks to come.

I located one of my Soldiers, informed him of the situation, and gave my first combat FRAGO, a quick-change order because of unexpected conditions. "SPC RT, you and Bennett, along with two Marines will make entry next door. Clear it and search for any weapon or threat. If you meet resistance, kill as many as you have to. Grab a foothold inside, and we will come for you. If there is no resistance, grab the family and bring them here. Do not hurt anyone unless you have to. Do you understand?"

"Yes."

"One more thing—search them for cell phones," I said, then frowned. This was getting complicated. "Screw it, I'll go with you. But it's your show." I didn't mind letting him take the lead. I trusted him completely. I didn't want anything to happen if their efforts met any resistance, so I followed them as another gun.

SPC RT moved smoothly, and we lined up, ready to breach the door. I took the lead, scanning the doorframe for booby traps—anything that might have been rigged in a hurry. I glanced at the point man and spoke loud enough for SPC RT, our number two, to hear. "When I open this door, give a brief pause. Then go." I'd picked up the technique in a tactical training course with LAPD, meant to prevent chaos in confined spaces.

Surprisingly, the door was unlocked. I turned the knob and gave it a push, then stepped aside as the stack moved in, each of us quick and precise. For a long, charged second, we flowed into the house. We methodically moved from room to room and encountered no resistance. The family members were gathered and searched, the men patted down while the women huddled nearby, tense but quiet.

After a quick frisk for weapons or phones, we escorted the men outside, instructing them to hop the wall. Even the elderly matriarch had to cross. She struggled, and we helped her over with as much care as we could. She offered us a few words of thanks—unexpected gratitude in a setting like this.

Inside, with a mixture of broken Arabic phrases and exaggerated hand signals, we managed to keep control of twenty-seven people. They stared at us, some afraid, others resigned. We were strangers in their home, wielding weapons and authority, trying to speak a language that wasn't ours—but somehow everything held.

Although we had no contact that night, my mind was racing a mile a minute with our newfound capabilities. In addition, I enjoyed working with ANGLICO. These guys were cool and calm. They were physically fit, competent, and willing to take risks to locate and kill the enemy.

After we returned, I approached CPT Rard to ask if I could stop by his office the next day. A massive individual who could seem intimidating at first, his leadership style was approachable and easy to work with. It was clear why so many respected him, and I felt a sense of ease working alongside the Devil Dogs under his command.

The next day, I showed up, uncertain of what kind of reception I'd get. I wasn't sure if CPT Rard would politely hear me out, then send me on my way, or if he'd actually take an interest.

To my surprise, he was just as enthusiastic about collaborating with Shadow 4 as we were with his unit. Encouraged, I pitched my idea to him—a concept inspired by a book I'd read years ago about a Recon Marine team in Vietnam, led by a Navy Corpsman, operating deep behind enemy lines. The team hadn't only scouted. They coordinated heavy ordnance strikes against enemy positions, acting as both eyes and muscle in the field. That image had stuck with me.

In a nod to the Marine Viper units, I suggested we name ourselves Task Force Viper. CPT Rard listened thoughtfully and soon drafted the concept of operations. The 506 signed off on it, but somewhere along the line, the task force took on a new name: Task Force Dark Eagle. It was official. We had our mission, our team, and a name that carried weight.

Doom on you, Hajji.

CHAPTER 8

SHUT UP, I'M GOING TO KILL THIS GUY

Inside the OP Hotel, Crows Nest Oct 2005.

I stood inside the room under white lights with the doorways covered in blankets to prevent light from escaping into the hallways. I was staring at the map with a small red light memorizing the satellite image by speaking aloud, "1,2,3 right, 1,2, left… hmm walk to Easy, look left and that should be it."

I was inside the OP Hotel's TOC. The fluorescent light added to the dramatic scene. My hair was matted down with sweat, although the air was rather cool, and some of the Soldiers and Marines wore their watch caps. I had my *shemagh* around my neck, more for comfort than tactical purposes. I hated a light wind going down my back, and the scarf prevented that. It was also a pillow for my head when I slept during ops.

Shortly after Christmas, we were attached to the 101st Airborne Division, 1/506 (1st Battalion, 506th Regiment).

LT AJ was a short, stocky man. He reminded me of a black-and-white photo of Audie Murphy, except LT AJ looked powerful. His right shoulder spoke volumes of the man. He was wearing a triple canopy: SPECIAL FORCES – AIRBORNE – RANGER tabs.

"You good?" LT AJ asked.

I didn't look at him. I kept studying the satellite imagery map. "Yeah, just making sure."

We had a large patrol going into Mula'ab. The 506 was not going to back away from this hornet's nest. There were four Marines (ANGLICO), four Snipers (Shadow 4), three scouts from recon PLT, the PSG SFC KL, and a CPT from Operations.

This was my plan. Although I was not comfortable with the size of the element, I also counted my blessings. More people equaled more guns.

This was my first patrol into Mula'ab, and I expected a fight. Call it a curse or good fortune, it didn't make a difference. I was convinced the entire city wanted us dead. If I could filter into the city and kill them in their own backyard, I was good with it. Sure, I carried my Catholic guilt, but I saved my tears for the Chaplain. Once I was past the wire, remorse was a flower that did not grow in my garden.

In January 2006, after only ten months, I received a promotion from E5 to E6, Staff Sergeant, a team leader. I knew I was changing, and SSG James Gilliand's words of encouragement fueled my spirit. Much was riding on me and on Shadow 4. I don't know if my Soldiers understood how we were viewed, but I saw it—some good and some not so good. Some stories were inspired by the rumor mill, but the common theme could not be denied. Shadow 4 was a respected element by both active duty and by our brothers in the National Guard.

We got to know the other sniper teams as we went on missions and met up in the chow hall. Staff Sergeant Luis Aguirre stood out, and not always in a good way. We had some things in common because our parents were Colombian, with South American roots that are a blend of Spanish and Indigenous ancestry, commonly referred to as Mestizo. Where my family was from the western side of the country and light skinned, Aguirre's family was Costeño, from the Caribbean side and dark skinned. He was short and stocky with curly black hair.

Aguirre was a good Soldier, the leader of a sniper team as I was, but he reminded me of that cousin or uncle in every family who shows up at holiday dinners and always has to be the best (or the worst) in the room. If someone has a health issue, his health was a thousand times worse. If someone mentions problems with their boss, his boss was the worst on the planet. We considered the source most of the time, but he could get on the guys' nerves, especially when tensions were high. Jokingly and lovingly, because he was our brother, we called him One-Up Aguirre.

Our team was gaining a reputation and respect. Some took steps to rein us in. Fortunately, their efforts were stifled by the leadership. Our success was difficult to argue. We were killing them and had suffered no casualties.

Looking around the room inside the TOC of OP Hotel, I felt energy in the room. Although I was a bit excited myself, I did not want anyone to notice and misunderstand. I wanted to be out there. All intelligence indicated this place was considered by the insurgents as relatively safe.

I wanted to put myself and the team in the middle of them to kill them and destroy their confidence, to destroy their will and completely shatter any thought that they could hide. I believed it would take a hell of a team to kill us. I was convinced we were that good.

We had learned the good and the bad from Shadow team. We needed to tap into boldness and audacity. I had wanted us to go into Mula'ab with a smaller team, but if this is what they wanted from the snipers, I was willing to take the risk of compromise. I needed the leadership to see us work and understand our process to build trust, confidence, and faith in my team and in me as a leader. Based on our recon assessment, I had chosen the house that would give us the best cover and vantagepoint to hold off an enemy attack.

Marine CPT Rard was a bear of a man, and his quietness told me all I needed to know. SSG G was relaxed and appeared to be in a good mood. As for SFC KL, his lean, weathered face was tight, but that's the way the man operated. He was ready.

I exited the TOC in the darkness and climbed the staircase to the crow's nest. My guys were inside the room with some of the Marines and scouts. A small heat lamp lit up their faces. Everyone appeared calm, no drama.

"Hey, guys." Their heads turned in my direction. "Start falling in on your gear and be ready to step out." I turned around and returned to the TOC.

LT AJ looked at me. "Whenever you are ready, sir."

I approached CPT Rard. "I'm ready, sir. Your guys and mine are upstairs mounting up."

"Thank you, Staff Sergeant." Marines never belittle the NCO. They always say *Sergeant*, never *Sarge* or *Sarg't*. In the Marine Corps, rank is not a suggestion—it's earned, and it's addressed properly. Unlike in the Army, where an E-5 and above might commonly be called "Sergeant" across multiple ranks, Marines don't do that. Every NCO is addressed by their exact rank—Sergeant, Staff Sergeant, Gunnery Sergeant, Master Sergeant—because each one carries its own weight, responsibility, and respect. You don't round it off, and you don't get casual with it.

God forbid a Marine forgets that rule. Getting someone's rank wrong shows a lack of attention to detail and respect, two things the Corps doesn't tolerate. Around Marines, you learn real quick: if they earned it, you say it right.

I pulled the blanket back and stepped into the dark, drafty hallway. The silence was thick, broken only by the occasional creak of gear or the shuffle of a boot. Around me, shadows moved—heavily armed Screaming Eagles from the 506th PIR, every one of them a ghost in the gloom, stationed around OP Hotel. Now and then, I could make out the outline of a Soldier manning his post, eyes trained into the blackness.

I found my gear and reluctantly pulled off my fleece and black watch cap, stashing them in my personally bought ruck. The cold bit into my skin, sharp and unfriendly. I ignored it, focusing on the ritual of checking my weapon. Pointing it safely toward the wall, I pulled back the bolt carrier, watching the casing settle into the chamber. I double-checked my optic, gave it two final turns to lock it into place, then flicked on my PEQ-2, looking for that faint, telltale green glow that told me it was ready. My body adjusted to the chill as I threw on my top, secured my plate carrier, and shrugged into my ruck. Its familiar weight settled along my spine. The Fritz helmet was cold against my scalp as I slipped it on, clinching it tightly for stability and securing my night observation device (NODs) with practiced care.

Once we assembled into the order of march (OOM), we filtered quietly into the street. Per the op order, we each took our assigned positions and, without a word, began our slow, deliberate entry into Mula'ab. The streets glistened with a thin layer of oily sewage, making them treacherous underfoot. As we moved, I kept my eyes low, avoiding the dark, slick puddles that dotted the path, every step calculated.

I ran the route through my mind, committing each turn to memory, until I reached the corner and saw the street I'd been waiting for—Easy Street, a four-lane stretch, bordered by a curbed median packed with dirt. The sight of it brought a strange mix of comfort and tension. From here, we were moving into the heart of the lion's den. In the next ten hours, these streets would be littered with dead bodies.

The point man took a knee. I was the second man, acting as navigator. Like a well-choreographed dance, we dropped like dominoes. I strode to the point man. He never looked at me. He kept his eyes to the front. As I approached, I touched his shoulder to let him know I was passing.

Slowly, he got up and followed me. We approached the corner together. I needed to first locate the house and try my best to avoid going over the median. The area was too dangerous to simply run across and hope for the best. We got here faster than I thought, so I had time.

To my left and across the street, I saw the house. It appeared bigger than the image from the Raven drone flight.

I decided to turn the corner and hug the wall to the next corner, as close to the house as I could. If nothing happened, that's the route everyone would take. They knew our tactics, that we liked to simply cross, but I had time and, besides, we were better trigger-pullers than the insurgents.

I explained my course of action to the point man and called the third man up to inform him as well. We moved along the wall, maintaining an interval. At the opportune time, I sprinted across the street, moving fast enough to look in front of me and still minimize my time in the street. As I covered my corner, I saw the point man repeat my actions. On cue, the rest of the element flowed to my position.

No one asked me about the way we crossed. We continued on. Two men at a time went over the wall while the rest of us pulled security. The gates opened, and we pushed in. Earlier, CPT Rard had said he wanted to try his pick-lock skills, and I agreed. I didn't want to knock and alert anyone to our presence. I watched the Marine officer and gentleman take a knee. Like something out of Hollywood, he worked on the lock.

It felt like a long time, so I approached and asked a question that I already knew the answer to. "How's it coming?"

CPT Rard had his helmet off, focused intently on the lock, hands moving with practiced precision. Every second felt like an eternity as he fiddled with it, the thought gnawing at me that someone could open that door from the inside at any moment. For a split second, I flashed back to a raid from my time as a police officer in Princeton, where the door had opened just as we approached. I didn't want a repeat of that here, not with everything on the line. On impulse, I gave the door a firm push, and it swung open. I glanced at CPT Rard, half apologetic for stealing his thunder. True to his professionalism, he simply stood up and gave a calm smile.

Wasting no time, I signaled to the nearest men, and we moved in. Like clockwork, the rest of the team followed, sweeping through the house. Within moments, we had rounded up every occupant and placed them in one room. They resisted as we ordered men and women to share a space. Even now, their culture insisted on separation.

I was astounded by their concern for modesty with guns trained on them, a reminder of how different our worlds were. A few choice words and a hard look from the end of my barrel quickly made it clear. I was here to keep control, not negotiate. Trust was their problem to earn, not mine.

Children's cries echoed through the tense silence. Women shouted at us, and the men muttered quiet defiance. Once we had full control, I left the decisions to the security element. With the Lieutenant now in the room, the atmosphere settled into an uneasy calm. I was certain he didn't appreciate my direct approach to "diplomacy," but my gut told me to stay on edge. Something was bound to happen.

I moved to the roof, setting up our primary and secondary positions and securing a 360-degree perimeter. The first SITREP came in over comms as the night stretched on, and we fortified our positions bit by bit. By dawn, we'd fallen into a steady rhythm, operating with mechanical efficiency despite the cold biting through our borrowed Iraqi fleece blankets. Huddled on the roof, layered against the chill, we waited, each of us locked in silent readiness. Every moment of peace held the potential for sudden chaos.

"Every moment of peace held the
potential for sudden chaos."

Settling behind the long gun, I scanned the street corner, watching as activity slowly picked up with each passing minute. Through the netting over our loophole, the harsh morning sun streamed directly into our position, casting sharp, slanting rays that illuminated the scene. My focus fixed on a young man, mid-twenties at most, wearing a black leather jacket cropped at the waist with blue stripes running down the sleeves. He moved with purpose, a shovel slung casually over his right shoulder as he walked toward the corner.

No way, I thought. He was waiting for someone, that much was clear. He faded in and out of my sight, only to reappear moments later, each time casting a longer shadow as the sun rose. Civilians had begun to filter into the street, going about their daily routines, but my attention stayed locked on him.

Every instinct in me said he was trouble, his whole presence radiating a quiet threat. I felt a surge, a readiness to end him for carrying that shovel. Something in my gut told me he was the enemy. I held off, knowing the rules of engagement wouldn't let me send him to hell without more reason.

Glancing down, I saw Bennett completely out, lost in sleep. No point waking him over what might amount to nothing. I waited, eyes steady, finger tense but disciplined on the trigger. My distance was 183 meters. Calmly, I adjusted my cam to 2 minus 1, then checked my natural point of aim, centering myself on the weapon, making it an extension of my breath.

The wind was slight, maybe 3-5 mph, half value with no visible mirage. The morning air was cool and crisp, sharpening my focus. I felt dialed in, as comfortable as you can be in a situation where seconds matter. The rifle rested on pillows to absorb the recoil perfectly into my shoulder, allowing me to reacquire the target instantly if a follow-up shot was needed. This setup kept the weapon anchored, steady, and controlled—eliminating the leap of the bipod during the firing process, ensuring I stayed locked on him.

The beat of my heart slowed to a thrum as I stayed ready, eye on the crosshairs, knowing that if the moment came, I'd need no more than a fraction of a second to take it.

The young man returned. An older man stepped out from a nearby shop to meet him, the two locking into a quiet exchange on the street. I watched every movement, every shift in body language. The young man reached into his jacket, pulling out a folded wad of paper, handing it over in a way that spoke of a significant payoff. The older man took it without hesitation. With that exchange, I flipped my safety off, the click quiet but unmistakable.

Stretched out by my feet, Bennett caught the sound immediately, mumbling, "Whaddya got?" from his half-asleep position, barely lifting his head.

I stayed focused, fixed on the scene below, scanning for any signs to confirm the threat. With both eyes open, I took in the entire street scene, pulling every detail of my surroundings into context—the noise, the shadows, the distance. I could feel the weight of the decision looming, even as my instincts screamed to pull the trigger.

Leaning into my Jersey drawl to keep the tone light and focused, I muttered, "Shuddup, I'm going to kill this guy." Tension surged as I watched the young man pivot and hand the older man the shovel. As I settled my crosshairs on him, the younger man turned his back and slipped into the crowd around the corner. *Dammit.*

My focus shifted back to the old man. He stepped off the curb, crouched, and began digging into the street, maybe four feet from the corner. He was positioned at an angle, visible enough to see the puffs of dust lifting with each heavy thud of dirt hitting the road. I slowly let the air out of my lungs, steadying myself, narrowing everything down to him. My left eye closed, right eye locked, my finger moved deeper into the trigger well.

Every nerve focused on that man, that moment. The street around him disappeared from my vision. A sharp *pop* echoed in the silence. The weapon recoiled, pressing into me, and I instantly chambered a second round. Through the scope, I watched the man's body jerk as if he'd taken a hard punch. His back was toward me as he staggered and slumped over onto his side, unmoving. The old man was dying or already dead.

My eyes immediately swept back to the crowd, searching for the younger man, the one in the black jacket with blue stripes. *Where are you?* I scoured the street through the crosshairs, breath tight and steady, teeth gritted. *Where are you, you fucking coward?*

I said, "Bennett, I just dropped an old man 183 meters to my 12. He accepted money from another source and began to dig near the middle of the street. He's wearing a black man-dress and white *shemagh*. He is down and not moving."

Got him.

Bennett was alert and moved to report it to the LT. Crying and wailing rose in the street. I stopped him. "Get on the gun and watch."

"Are we going to move?" he asked.

"Not yet. Let me report it to the LT." I got off the gun, and he quickly filled the space. As I reached the door jamb leading down the stairs, I encountered LT AJ and reported my actions. He quickly absorbed the information and went back downstairs.

"SSG Martinez, there is a crowd forming, and they are coming up the street," Bennett said.

"How many?" I visually checked my security ring. I called the LT on the radio and informed him of the crowd, then said, "Bennett, keep watching… How many?"

"Forty. There's a lot of screaming going on."

I could hear it. I looked at CPT Rard who was getting his team ready.

Bennett said, "SSG Martinez, yeah, they are rolling in a bus and blocking off the road. There's a man near the bus, and he is on a phone. It looks like he is organizing the others. He's pointed twice in our direction. I also observe a group of military-aged males on different rooftops. They are scanning their areas. Many of them have weapons."

"If you see a man with a black waist-high jacket and blue stripes running from the shoulder to the wrist on each arm, kill him. He provided the money."

"Roger."

CPT Rard said, "SSG Martinez, we can have some fast movers here real soon. I'm calling brigade to see if they will allow us to drop an LMAV."

"Great, sir." I wasn't too concerned with the populace, although I was surprised by the bus tactic. It was a clever maneuver.

Bennett said, "SSG Martinez, they are moving around in groups of five and knocking on gates. The guy with the cell phone seems to be guiding everyone. He is almost completely covered behind the bus."

"If you are certain, you can kill him. Go ahead."

"He just pointed in our direction."

"Kill him."

His weapon rocked Bennett back. He followed up with a second shot. "SSG Martinez, I shot him twice. The first shot hit him under the arm as he pointed, and it spun him. I shot him again, and he fell against the bus. He's not moving."

"Good job. Stay alert."

I saw a figure move in the doorway. LT AJ. I moved to the doorway and reported Bennett's actions.

AJ appeared extremely focused. I informed him, "Sir, we have fast movers approaching."

"Roger." He turned around and went downstairs.

I located CPT Rard, who was talking on the radio. I could hear the jets in the air. My ICOM radio was alive with radio traffic. Someone was knocking on the gate.

I got on the radio. "Get the homeowner. Tell him not to betray us, or we will kill him. Have him speak to the guests, but do not let them inside the courtyard."

A few minutes later, the radio came on. "Shadow 4, they are gone. I don't know what he told them, but they are gone."

"Roger." *Had the homeowner betrayed us?* I returned my attention to CPT Rard. "Hey, sir, how is it going with the fast movers?"

CPT Rard spoke as if everything were cool as ice. "They have denied us the LMAV, and I am requesting to open with guns.

Shots rang out and impacted our southern wall. One of the Soldiers looked through a portal in the wall and returned fire. As I watched our perimeter, SFC KL appeared, and I shouted, "We're good!"

Without acknowledgement, he descended the staircase. Again, the southern corner was attacked. The Currahee Soldiers responded with a heavy superior fire.

Bennett said, "SSG Martinez, there are more people in the street to my 12. I see weapons."

"Kill as many as you can. Maintain the 360!" I ran low to the southern corner and peered into the street through a hole in our wall. A few bodies lay in the street. I saw the crowd, and I saw the weapons. I searched the crowd for the young man I felt was responsible for it all. I wanted him dead. I wanted to kill him.

As I returned to CPT Rard, I kept hearing a repeated phrase on the radio, "Bent rod."

Moving closer, I said, "Sir, I know you are pretty busy, but how the fuck do you bend a rod on multimillion dollar plane?"

With the patience of an instructor dealing with an ignorant man, CPT Rard stopped speaking on the radio and cupped the lower end to spare a greater audience of my stupidity.

"Staff Sergeant, you don't. It's a phrase indicating there is a glitch in the system. We don't have guns. So, he is going to do a show of force."

He requested permission to drop a GBU-12—a 500-pound laser-guided bomb. Under the ROE, that level of firepower required approval from the Division CG (Commanding General), General Kelly. The request was denied.

He adjusted immediately and called in a gun run from "Fires" Dealer 11. "Fires" in the military means using weapons that hit the enemy from a distance to support the fight. Think of it like this: if infantry are the ones closing in and fighting face-to-face, "fires" are everything backing them up from afar dropping bombs, shooting artillery, launching rockets, or firing from aircraft to destroy or suppress the enemy before Marines or Soldiers have to get too close. "Fires" is the King of Battle, the military's way of reaching out and hitting the enemy hard before you ever have to shake his hand.

In this fight, "Fires" were supposed to be under his control on the ground. Instead, the gun run was abruptly aborted. No explanation. No warning. CPT Rard pushed back—same result. Aborted again.

He shifted to show-of-force runs, maintaining pressure while trying to figure out who was interfering with his control of the fight. Over the radio with USMC CPT Murray, the frustration was clear—what the hell is going on? It didn't take long to piece together that they had a "bent rod." Someone higher up was reaching into the engagement and making decisions without situational awareness.

CPT Rard scrapped the plan. He brought in the Cobras to regain control. With that, he returned to his radio.

I looked at his Marines. They ignored me and continued watching their lanes.

At the time, none of it made sense.

Years later, Major Rard—having since been promoted—explained what had happened. Division had been the one aborting the gun runs.

They believed the Marines were engaging the Mala'ab militia they were trying to stand up, and from their vantage point, they intervened.

But from the ground, it was something else entirely.

He would later recognize that requesting the bomb had drawn Division's attention into the fight. That decision brought higher headquarters into a situation they didn't fully understand, and they acted on incomplete information.

That became the source of the concern. Not the enemy—but leaders making battlefield decisions who were not on the battlefield.

I could hear a SAW M249 gun spewing death from our northern position on the roof. SGT PH was getting some.

"Bennett, put the long gun down and start killing them with your A4."

I wanted to jump in and throw down, but I had never fought with nearly half of these men. I was not concerned with Shadow 4 and ANGLICO, but the Screaming Eagles were new to the fight. I stayed in the middle to make sure everyone was doing his job.

LT AJ appeared at the doorway to inform me QRF was enroute with Bradleys.

"Roger, sir." What else could I say? I went to the southern corner to locate targets and shoot.

A loud roar came over our position. I didn't see it, but judging by the street's reaction, they did. From the corner of my eye, I saw chafe dropping. People were running everywhere. Whatever they thought it was, they didn't like it. Our perimeter was engaging from the north and southeast sides. The west side of the perimeter maintained its sector. I heard no rifle fire. They were a disciplined lot.

The Screaming Eagles were living up to their name. From my vantagepoint, CPT Rard was getting some. He had a good angle, his rate of fire slow and deliberate, and he was hitting. Before the day would finish, this Harrier pilot would send five to martyrdom.

The level of fire was escalating. The majority of the fire coming from the street was poorly aimed. We were seizing the initiative, and they were reacting to us. A couple of times, men joined us on the roof to get into the fight. When their good judgment and discipline came back, they would return to their position inside. That was the only way to avoid a cluster fuck on the roof.

From the north, I could hear Bradleys spitting hate on that street. They looked enormous below us. Three of them entered the street between us and the neighbor's house. Scanning the street for more enemies, I waited for the word to head down, but it never came. I watched our security element and the leadership exit the courtyard and begin to mount up. Without any further instruction, the team systematically went downstairs. By the time it was my turn, it was just me and the ANGLICO Marines.

Stepping out onto the street, I saw the fourth Bradley out on Easy Street receiving SAF (small arms fire). I put out a call over the radio to the LT, asking to have the TC drop the ramp so we could come running in. A short time later, the LT came on the radio and said, "TC cannot drop the ramp. He wants someone to run and knock on the rear hatch, and then he will let you in."

That request required a Soldier or Marine to run out completely exposed to enemy fire, to stand there and knock on the steel hatch, then wait until someone opened it. Only then could we move into the Bradley fighting vehicle.

Truth be told, several things entered my mind. (1) I got the Marines into this fix. If one of them gets hurt or killed, this is my fault. (2) Where is the Star Trek guy in the red shirt when you need him? (3) I'm going.

I yelled, "COVER ME! *I have to knock on the hatch.*"

Before I could fully process my own decision, my legs were already propelling me forward, pounding against the ground. I barely felt it. *Get to the hatch. Get to the hatch! Get to the hatch!* The words looped through my mind as I sprinted, my arms tucked tightly around my head and shoulders, rifle clutched in my hands like it might shield me from the rounds ripping the air. The world shrank to the distance between me and the ramp.

I pounded the door. The lock rattled. I jumped back as the one-ton ramp slammed to the ground. Every instinct screamed to rush inside, to shout for the others to follow, but discipline held me firm. I was an NCO. Instead, I took a knee outside the ramp and faced the street.

CHAPTER 9

JOKER 4

Shadow 4 with Shadow 6, SSG James Gilliland and SPC M. Ferguson from the 1/506 RECON PLT December 2005

Raising my rifle, I fired downrange, rounds cracking from the barrel as insurgents darted out from cover, attempting to close the distance.

The Bushmaster's cannon opened up, filling the street with its heavy, guttural roar. Its barrel spat round after round, ripping through brick, glass, metal, and flesh—anything unlucky enough to be in its path. The street lit up with muzzle flashes and choked with the smell of gunfire, the air dense with heat and adrenaline.

As we moved to mount the Bradley, our backs were completely exposed, boots pounding, adrenaline spiking—and that's when three enemy fighters came hard, trying to buttonhook the ramp and pour fire straight into the Marines and Soldiers before they could get inside. We were open targets to an enemy already maneuvering into position to kill us.

Through the fog of war, CPT Rard appeared, the last to step onto the ramp, and I was right behind him, both of us covered in dust and sweat. Suddenly, he stopped, stepping back onto the ramp, raising his weapon with a calm, brutal precision. As I reached him, I saw his target—a man in a black *dishdasha*, face hidden beneath a headscarf, charging us with an AK-47 raised at his hip.

Time froze. The man's eyes locked on us, his weapon swinging up. CPT Rard saw it unfolding a split second before it could turn fatal.

Rard didn't hesitate.

His first shot cracked through the chaos and hit one man square in the chest, snapping him backward and dropping him where he stood. The second came into his sights almost immediately. Rard fired again, rounds tearing into the man's hip, spinning him off balance and slamming him to the ground. The third saw both men go down and broke, turning and running for his life.

It was fast. Brutal. Up close. The kind of violence that doesn't leave time to think—only to act. A few seconds slower, and we would've been cut down with our backs turned, with some of us never reaching the inside of the vehicle.

CPT Rard lowered his rifle, his expression grim, focused. As he took his seat he muttered, "That's five." No boast, no hint of celebration—just a simple acknowledgment of deadly skill honed by experience.

The ramp closed, and I realized how close we had been to death at that last minute. Once more, cool heads prevailed, and marksmanship was dead on. Mula'ab had just suffered the wrath of the Eagle, Globe, Anchor, and Screaming Eagles.

Able Company of 2/69 3ID heard our call and responded by sending us their BFVs. We returned to base. When I stepped out of the BFV, Soldiers were high fiving each other. I smiled. They were celebrating life. 506 had just announced itself in Mula'ab.

To my right, a young LT named Johnny Bravo said, "I knew it had to be you, Shadow 4. Good job." Assigned to the 2/69 3rd IN Mech, he had answered the call to arms immediately after 9/11.

"Thank you, sir. Thanks for coming to get us." I walked away so he wouldn't see me shaking.

Years later, I ran into CPT Rard in San Diego. Over rounds of beers and memories, we talked about that day, revisiting details of the mission with the clarity only time and distance could bring. At one point, he casually mentioned that our small unit had held off more than three hundred fighters. Those numbers stopped me in my tracks. They marked a collective effort—the resolve, skill, and unbreakable mindset shared by the Marines and Soldiers who carried the fight. That truth humbles me to this day.

Shadow 4 becomes Joker 4

On my last day of rotation at OP Hotel, I got the official word: Shadow Team had been stood down. Joker Team from the 1/506th of the 101st Airborne (Air Assault) was taking over our AO. For all of us in Shadow 4, this news came with a mix of apprehension and pride. We knew the 506th well. It was "big Army," a reformed unit with a legacy made famous by Stephen Ambrose's *Band of Brothers.*

That legacy was impressive, but we worried it might come with egos that would make them look at us, a National Guard team, and say, "Thanks, but no thanks."

We arrived just after nightfall, filthy, exhausted, unshaven, and worn out from the rotation. Despite the exhaustion etched on every face, I asked my team to hold off on getting settled, so we could introduce ourselves to Joker Team right away. From the chow hall, we walked over to Full Metal Jacket, each step a reminder of how far we'd come.

A few months ago, we'd barely known the city, our roles, or what we were capable of in combat. We'd almost drowned before we had a chance to prove ourselves. But now, Shadow 4 has made a name on Corregidor. Known for our discipline, precision, and dedication to the art of sniping, we'd gained respect as professionals, even killers—though not without controversy.

Our team's modified uniforms set us apart. Some questioned our morals, not out of reason but out of envy and a misunderstanding of our methods. "Hey, Killer," someone greeted us as we passed, a label we'd earned, for better or worse. Rumors swirled, not always kind, but they reflected the impact we'd made. Shadow 4 had come a long way, though we were only halfway through this deployment.

When I opened the door, Gilliland was standing near his cot. He was addressing Joker Team—made up of ten Soldiers, three team leaders, and one section leader—each looking like they'd been through their own battles. From the expression on Gilliland's face, he hadn't expected us to show up.

"Harry, this is SSG Ken Cooke, section leader of Joker Team," Gilliland said, introducing us.

I extended my hand, offering a firm shake. "Harry Martinez, team leader of Shadow 4."

Around me, some faces were smiling, others guarded. One by one, they introduced themselves, and I reciprocated, introducing myself and my team. The tension was palpable. I felt once again we needed to prove ourselves. We had our reputation, but that didn't guarantee anything.

"Harry, why don't you and your guys clean up and come back? I'll fill you in on what's going on."

I took that as my cue to leave, offering a nod in return. Something about the moment felt significant—the beginning of a new chapter, but with uncertainty lingering in the air.

Walking back, I felt a mix of hope and apprehension. This could work. It had to work. At the same time, I had to make sure TF Dark Eagle and Joker Team didn't clash, and I had to keep ANGLICO in the mix. They were indispensable for mission success.

My mind was racing at a hundred miles per hour. I was hungry to keep fighting, to leverage every asset at my disposal—especially if ANGLICO were still willing to locate, engage, and kill the enemy.

Nearly seven months into the deployment, it became clear that not every Soldier, even among our leaders, was cut from the same cloth. Some were going through the motions, hoping to pass the buck to the next unit. The unwritten rule seemed to be: if they don't bother us, we won't bother them. Even within Shadow, this apathy had begun to seep in.

Before Shadow left for CONUS, a team leader had been relieved, and in my opinion, another should've been. The team leader who replaced Shadow 1 impressed me with his ability to move within the shadows, as the old saying goes. Sometimes the long way was the hard way, but it made sense in combat. When lives were on the line, there was no room for shortcuts.

If SPC PR had stayed long enough, he would've excelled with the team. LL, on the other hand, was blunt and often remarked, "I'm about to do something dumb." True to his word, LL rarely let us down. I was taken aback by his sheer audacity. He's the guy who single-handedly captured an enemy sniper and brought him back for interrogation. Both leaders had their strengths, but for me, their shortcomings often outweighed their virtues.

Sure, we all made mistakes—I'd had my own. Once, I fired my weapon unintentionally, known as a negligent discharge, which could've cost me dearly. By all rights, I should have been relieved. But no team was perfect, just as no family was. I was forgiven for my error, and SSG Gilliland's words stuck with me: "Harry, even world-class shooters have accidental discharges (ADs). Just mind your manners." He slapped my hands.

That second chance meant everything to me. I respected SSG Gilliland more than I could express. My loyalty to him was unwavering, and I knew, deep down, that our team had made history. 3rd ID 2/69 Shadow Team had become legendary with over 260 confirmed kills in just five months. SSG Gilliland had secured one of the longest sniper kills on record, using an M24 SWS (Sniper Weapon System) to take down an enemy sniper from 1250 meters. We were snipers, and snipers measured their success by the enemies they hunted and destroyed.

After dinner one evening, one of the guys was talking about an old sports injury, and SSG Aguirre interrupted him with a story of his own injury that was far worse. I wasn't in the mood for it, so I decided I'd beat him at his own game. I interrupted Aguirre with my own story and—sure enough—he came back at me. So, I pulled up another. I'm not sure what my objective was, but the evening ended with everyone irritated. That was the last time I tried something like that, but I have to admit the guy was getting on my nerves.

As Shadow began to stand down and Joker Team took over, I was determined to keep ANGLICO involved in future operations and to bring the Currahee BN scouts into the fold. Shadow team incorporated selected scouts, and only three were ever selected to accompany us. I never questioned SSG Gilliland's decision on that. It wasn't my place.

When I returned to Joker Team, I could feel the energy in the room. These men were eager to operate, to make their mark as snipers. Their section leader left a strong impression on me as a combat leader. Over time, I would hear some grumblings and complaints about the team, but I paid them no heed.

Their section leader, SSG Cooke was a professional—highly capable and determined to get the job done. Of average height, he was a quiet-spoken, cerebral man with piercing blue eyes who was quite capable of violence. He would earn my trust and loyalty, a true leader among savages.

By the time I returned to the States, I knew I had been exposed to some of the finest leadership I'd ever encountered in my military career. I repeated that often, even if they grew tired of hearing it. I want them to know I hold them responsible for my success and survival in Ramadi. Without their leadership and guidance, I wouldn't be here today.

Abandoned

The 506 had transformed part of their living quarters into a fully functional Scout/Sniper Operations Center (SSOC), a hub of real-time situational awareness that brought a new level of coordination to the battlefield. Inside, an overhead projector cast tactical data onto the wall, while a Single Channel Ground and Airborne Radio System (SINCGARS) radio crackled with updates and commands from across the AO.

A long table, surrounded by a few chairs, served as the nerve center. A massive satellite imagery map dominated one wall. Terrain detail was outlined on that map. An erasure board stood nearby, filled with hastily scrawled notes, grids, and priorities. Other essential tools lay scattered around the room—extra batteries, notepads, and spare comms gear—all organized with precision.

At first, I thought it was overkill—a lot of bells and whistles for what was ultimately a chaotic and unpredictable fight. But as the days wore on, I saw the brilliance in its design. The SSOC was a force multiplier. It gave us something invaluable: clarity.

The ability to monitor and adapt in real time, to coordinate across teams, and to visualize the battlefield on a scale that a single Soldier couldn't comprehend in the heat of combat—this was empowerment. The satellite imagery allowed us to anticipate movements. The radios kept us connected, and the combined tools enabled us to turn raw data into actionable decisions. The SSOC was the heartbeat of the mission, a place where strategy came to life and where success felt achievable in the darkest moments.

We planned and refined our missions with exactitude. Working alongside the Joker 506 snipers was a shift from what I was used to—one that cultivated unparalleled professionalism. Every detail mattered, and every movement was deliberate. What initially felt rigid became the framework for seamless execution.

When LT AJ, the Platoon Leader (PL), led a briefing, I walked away empowered. He had a way of breaking down complex operations into clear, actionable steps, leaving no room for doubt or miscommunication. His methods made the command hub a place of simplicity and purpose. You left the table feeling confident. The SSOC turned intent into action, ideas into strategy, and elements into cohesion. They set a new standard.

Consequently, little changed in the overall structure of how Joker and Black Sheep (506 scouts) planned and executed missions. The integration of their disciplined approach sharpened the process.

Mission briefs were concise but thorough, emphasizing adaptability and precision.

Up to this point, I had been working closely with SGT TOM and his scouts, but tonight I was assigned to SSG JJ and his team. SSG JJ carried himself with confidence, the kind of Soldier who talked a good game and seemed ready for anything. His "tower of power"—the Airborne and Ranger tabs on his shoulder—symbolized his training and hinted at a leader capable of shouldering the burdens of war. He exuded an attitude that said he could handle whatever came his way.

Combat leadership during the Global War on Terror (GWOT) from 2005 to 2006 was uniquely demanding, particularly in the fight against an insurgency. Leaders were more than tacticians and planners. They were the emotional and psychological anchors for their teams in an environment defined by anarchy, uncertainty, and ruthless pressure. The emotional toll on this leadership came in their understanding of the human cost of war. Managing the invisible toll it exacted on every Soldier was a heavier burden than any physical responsibility.

"The emotional toll on this leadership came in their understanding of the human cost of war."

The insurgency in Iraq posed a unique set of challenges for leaders. Unlike traditional warfare where objectives and enemies were clearly defined, insurgent tactics thrived on ambiguity. Leaders had to navigate a battlefield where the enemy blended seamlessly with the civilian population, where every vehicle might be a VBIED, and every bystander a potential scout. This demanded extraordinary adaptability and mental resilience from leaders who were expected to make split-second decisions in morally and tactically gray areas. They had to navigate in the fog of war, where transparency was fleeting, death was always nearby, and decisions carried immediate and irreversible consequences. This demanded more than tactics. It required the ability to inspire trust, maintain focus, and make critical judgments in what many referred to as the graduate level of war—the brutal, unrelenting complexity of the Global War on Terrorism.

War erodes the strongest among us. As committed and skillful as he was, SSG JJ reached a breaking point from the cumulative effect of constant danger, mental strain, and the weight of responsibility. Watching his decline, I felt a mix of detachment and empathy. Detachment, because in war, survival demands a shield between your emotions and reality. No amount of training, tabs, or accolades makes anyone immune to the soul-crushing intensity of combat. His struggles were not a sign of weakness. They were a stark reminder of how deeply war changes every Soldier.

In the days leading up to the mission, LT AJ approached me with concerns about SSG JJ. He believed my experience and presence might provide some stability and encouragement to SSG JJ and his team. This was not an unusual request. Combat leaders are often tasked with supporting one another in ways that go beyond tactical execution.

This mutual support was a testament to the camaraderie and unity that underpinned our leadership in this type of war. It required professional competence and the ability to recognize when someone was faltering and intervene before the cracks became catastrophic. The mission ahead was about reinforcing that sense of trust and stability among Soldiers as much as it was about combat. Supporting SSG JJ was an acknowledgment that in war, even the most potent leaders sometimes need someone to lean on.

I informed SSG JJ that my team would attach itself to his for the next mission. The conversation was straightforward. I didn't want to undermine his authority or draw attention to the concerns. Instead, I framed it as mutual support, a fundamental principle of combat leadership. I emphasized that my team would supplement his efforts, not overshadow them, and that we were there to share the burden and ensure the success of the mission.

The mission unfolded along a desolate two-lane road in the unforgiving expanse of Anbar Province, 2006. The road was a vital Main Supply Route (MSR), cutting through the arid landscape like a scar. The desert stretched endlessly on either side, the sand shimmering in the oppressive heat, which brought up the kind of mirage that snipers used to aid their shot. Occasional clusters of square buildings formed the edges of hostile neighborhoods. The air was heavy with tension, and every movement carried a sense of unseen danger.

A Marine survey team was filling in potholes that threatened the flow of military convoys. On the cracked pavement, a young female Marine stood in the middle of the road, scanning the surrounding area with a measured gaze. Soldiers from the 506 watched her every move with a mixture of curiosity and unease. Another Marine approached her, and the two exchanged words in the open, standing exposed on the MSR without cover.

As the male Marine finished their brief conversation and headed back to his vehicle, the young woman lingered, pivoting to face the shadowy, unpredictable neighborhood that bordered the road.

A shot shattered the calm.

The sound echoed across the barren terrain, a sharp, definitive crack. She froze momentarily, her hands lifting slightly into the air. Then, she collapsed. The desert swallowed the noise as quickly as it had come. She hit the ground, crumpling in on herself like a broken doll, her helmet skewed. A single, precise round had found its mark, striking her beneath the brim of her helmet, directly in the face. Death claimed her before her body hit the asphalt, a quiet reminder of how fragile life is in the unforgiving theater of war.

The checkpoint was taking sporadic fire from within the neighborhood. Enough was enough. Joker elements were tasked to locate the enemy sniper and kill insurgents attempting to commit a hostile act against the CF, the checkpoint, or a CF patrol, according to the existing ROE for Ramadi.

Joker 3 and 4 poured over the maps. We studied our Raven UAV flights and all SIG ACTs about our final destination. In addition, we did a foot recon of the area and selected two potential hides, a primary and a secondary. We continued to monitor the area. On this particular night, when we inserted into the area, I was seventy-two hours away from leave.

As we moved out, I felt we were poor in our movement. We bunched up together. A couple of times, we had moments of indecisiveness.

I looked behind me and counted my guys. We weren't going far, and the streetlights caused a white flash with my NODs. I had to lift them to look over my shoulder. Bennett was on Tail-end Charlie performing his Belfast Sweep. SPC RT was positioned between us, scanning the buildings and searching from left to right. We were fine.

I said, "I'm going to move up. Keep an eye on each other." SPC RT acknowledged me, and I moved almost directly beneath a light. *Stay out of the light!* I slid into the shadows.

A moment later, I spoke to SSG JJ. "Hey, SSG, what's up?"

He replied, "I don't remember the house looking like this. I thought it was better positioned." His head was moving in all directions.

"We can always go to our secondary. Let's move and check it out."

He looked at me like I had six heads. Hesitating, he said, "All right, let's go."

"Great. Bennett takes point," I ordered.

"Fine."

I returned to my place in the formation and told SPC RT to tell Bennett to lead us to our second choice.

Bennett nodded and stepped off. I looked back over my shoulder and told SGT TH, "We're moving."

From the outside, the front of the house looked better than the primary. Bennett took a knee and looked down at the street. I eased my pack off and put my weapon down. SPC RT followed and placed his back to the wall. I climbed onto his cupped hands, and he lifted me up. My weapon passed to me as soon as I got on top, and I dropped to the other side.

I looked around and waited perhaps thirty seconds before moving. At the gate, I watched for anything unusual, then opened the gate. Everyone moved in. For a brief moment, we surveyed our environment, looking at two structures. One was a small one-floor unit, and the other dwelling appeared to be a two-story structure with many rooms.

I said, "Let's hit the big house first."

Security moved into the shadows, and we stacked on the door. It was unlocked, which seemed unusual. The house was empty. We went upstairs and through a door that led outside again. We found two more rooms, and I searched one of them with two other Soldiers. The room was long and rectangular with a lot of furniture and random household items stored there. I spotted a second door and signaled to the others.

One of them had the door in his sights.

I moved to his front, avoiding his barrel. When I opened the door, I immediately sensed something wrong. I was looking outside. I didn't move. The two men behind me stopped short.

"Stop! There's no floor. Be cool." We backed up, and I looked for options. I went downstairs and linked up with Patrol Leader.

"Hey, man, this house looked good from the outside, but the second floor is totally fucked. In the daytime, our ass is hanging out in the wind. It's a house under construction. What's in that small structure?"

'I don't know."

"Let's go check it out." Again, I used a tone that could only be interpreted as an order.

After we searched the house, we decided to stay. The security team would set up in the small structure, and the snipers would move to the larger structure. SSG JJ added that he and the RTO would remain in the larger house on the ground floor.

We checked our radios, and the security element remained behind. SSG JJ appeared a bit rattled, but I said nothing to him. SPC RT followed me to the second floor and helped me build a shooting platform using the materials inside. We found a perfect window, giving us a linear depth field of view.

SSG JJ would periodically come to our door as we settled in and established our radio checks. I was getting concerned about his behavior. He appeared to be distracted and preoccupied and loaded with anxiety.

I said to him, "Brother, you good? Look, this might work. I got an unobstructed view, and this was the road they used to shoot at the checkpoint. We'll be fine. Uncomfortable but fine. Don't come up here anymore, and make sure the guys implement a rest plan." He didn't respond, but I felt it was better to be forthright.

Nearly an hour later, SSG JJ called me on the radio and reported he had an urgent message. I said, "Joker 4, Send it."

A voice on the radio said, "Joker 4, Currahee 6 reports intercepting a series of messages near our AO. Break." *Cool Man, you gotta love technology.* "Prophets report they have located an American unit inside the neighborhood. Break."

Oh, shit.

The radio operator added, "I asked them to double-check and see if there are friendlies in the area. Break." *Christ, is he doing that for effect?* "Currahee x-ray reports negative on friendlies."

I replied, "Roger, I'm coming down to you. Joker 4 out."

Well, this changes everything.

In the darkness, I saw the RTO drinking a bottle of water while monitoring the radio. SSG JJ was pacing. SIGINT (Signal Intelligence) had intercepted a message reporting the enemy had observed an American patrol, and the enemy was amassing forces preparing to attack them. Since there were no other friendly elements in the area, it had to be us.

I wanted to stay. After all, I was an infantryman. We were loaded for bear, and I felt confident we could take on anything they threw at us. I told SSG JJ, "Hey, man, let's do this. If it's us, let's release the family, consolidate our guys and build up our fighting positions. Our job is to get them to come out. If they want to come out in force, we have some laws, 203, and hand grenades. This house looks like it can take some heavy RPGs. I think we are okay. We just won't be sniping anyone."

His face said it all. "I don't think so. I think we should go."

I pressed him. "We have two trucks and a tank 300 meters behind us. If it gets really bad, we'll request QRF and have the tank flex, or someone can flex from the checkpoint."

Another NCO spoke from the radio in his calm southern drawl—which I hated to hear at that moment. "Hey, I think we should break down and call it a night. This is not good."

I was getting pissed, but SSG JJ called it and ordered us to break down the position. There was no sense in arguing. I went upstairs and started to move everything back. I was cursing SSG JJ and that other NCO from the radio. The other Soldiers did not need to hear his doubts. As I understand combat leadership, that could cause the Soldiers to lose confidence.

SSG JJ reported our request for extraction, and they replied they were enroute for exfil. We continued to break down our positions while trying to remain unknown to the enemy.

We gathered in the courtyard and exited through a hole in the wall. As we approached the main MSR, we linked up with a 113 and returned to base.

As we dismounted, I told my guys to return to their rooms and get some rest. SSG JJ and I proceeded to the BN TOC. Waiting for us at the top of the stairs was a very pissed off LT AJ. He asked us to follow him to the conference room.

This is not going to be good.

We sat at the table for our debrief. Once we were done, LT AJ asked me for my opinion on the situation. I told him the truth.

"We should have remained in place, considering all the assets we had in close proximity. I am prepared to go out agai…"

His tone changed, seething with contained anger. "Goddam right you are. Tomorrow, you will return and locate another position for the same purpose. Are we clear?"

I looked at SSG JJ, and he did not look happy. It was as if he had just received a death sentence. At that moment, I decided I wanted to assert some control: "SSG JJ, I want to address the NCOs before we leave. I will see you and them later at the SSOC." I was not requesting anything. It was a standing order. Surprisingly, SSG JJ agreed.

The next afternoon, we gathered in the SSOC and the NCO from the radio was present. My remarks were intended for his and SSG JJ's ears.

I told them, "Next time any of you have an opinion on a mission, I want us to gather together and discuss it. I do not want to hear doubt over the radio. Our non-rates look to us for leadership. We cannot instill a sense of dread during a mission. Do you understand? No matter how bad it looks, we will always appear in control. Next time, stay off the fucking radio and keep the negative to yourself."

That NCO appeared resentful of the remarks. I could not care less. My experience convinced me I was right, and I was not going to budge left or right on the subject.

No one disagreed or added any remarks. "Good. SSG JJ and I will coordinate, and you will prepare the trip ticket for BN. I will see you all at 0200. Are there any questions, problems, or complaints?"

The NCO jumped off the table. His manners appeared belligerent. SSG JJ neither endorsed my comments nor disregarded them, and that was good enough for me.

Bennett and I showed up as planned. Everyone was ready. I said little. Since we had already done numerous missions, I saw no need to be on top of the guys. At approximately 0200, we staged outside and waited for our transportation. We were inserted by HMMWVs with an open back. During the fifteen-minute ride, I was freezing my ass off. I was grateful once we stopped.

We entered the neighborhood and selected a house. Following all SOP's this time, I knocked. A light went on inside, and someone spoke through the door.

I called out, "Open the fucking door, Amerikee Jundee (American Soldier)."

I heard the door unlock and pushed myself in. I grabbed the man who was in his late forties and shoved the barrel of my pistol under his chin.

CHAPTER 10

DARK WATERS

Joker 4 prepped for a recon patrol with 1/506 Sniper Section leader, Winter of 2006. Ramadi, Iraq

"Shut the fuck up, or I will kill you!" I pushed the man into a counter, and he dropped to his knees.

He was whining, and I was straining to listen for any anomaly. "Shut up!" He went silent.

Behind me, the men flowed in, and yelps of surprise followed. As the team searched, we found a lot of Iraqi dinars.

"No Ali Baba!" the man yelled at me.

"Shut up!"

He became quiet and submissive. That worked for me. I kept my pistol on him.

SSG JJ walked in and said, "Shit, man, you sounded serious about killing him."

SSG JJ continued to oversee all radio communications while I coordinated with the scouts. We occupied two positions in adjacent rooms on the second floor. We had the family move into a separate room. The man's sons were of military age and appeared muscular and tall. These boys were definitely into sports.

Since we had a terp (interpreter), I interrogated the boys. At the time, this was permissible, but later the rule on interrogation changed. I decided to interview the oldest one.

SSG JJ followed me into the room. To my right sat the terp. I invited the Iraqi young man to sit. I introduced myself and told him words to the effect that I am an American sniper. My job is to kill anyone who interferes with Coalition Forces. Unfortunately, I believe that sometimes Iraqis were committing acts that appear hostile, and because of the current situation, we kill them. Therefore, I wanted to know what we should look for to avoid killing innocent Iraqis.

The boy looked at me and remained silent.

I tried another angle with the same theme. "Look, I know we appear hostile and angry. Try to look at it from my point of view. I am here to kill anyone who appears hostile. I want to avoid killing innocent people. For example, let's say your brother is playing soccer out in the street and does something that catches my attention. I might feel that his actions are intended to signal an attack, but it's an innocent gesture. I don't know that, so I kill him. Don't you want me to avoid killing your brothers and your friends?"

The Iraqi looked up. With a sense of resignation, he murmured, "If Allah wishes it so, then it is Allah's will."

I pointed at him. "You are wrong. Tell me how I can identify the Ali Baba (bad guys). Does he wear certain clothes? Does he make certain gestures? Does he roll his pants a certain way? How do two AQ members greet or identify each other without ever meeting each other before? There must be a signal."

Again, he looked up. "I do not know. I do not know these people."

I stared at him. "I think you want to tell me, but you are afraid. I believe you want to help me. Please try to understand that your silence gives these people power. By helping me, your brother and friends remain alive. If you choose to say nothing, it is not Allah's will but yours, and the death and blood are on your hands." I made a gesture for him to leave.

As he got up and walked to the door, I stepped in front of him, my nose two inches from his. "One more thing, before I leave this house, I will kill someone. If his death was unnecessary and could have been prevented, know that it is your fault for being silent."

He turned and walked away.

When I sat down, I noticed a cup of tea had been set aside for me. I picked it up and drank it slowly. SSG JJ sat down next to me and said, "Hey, man, all your talk about killing is disturbing me."

By now, I had been through nine months of killing people in Ramadi. Although a core terror stayed with me every second I was there, I had accepted my death and a part of me felt fearless with no remorse. My heart was under dark waters—cold, steady, and resolved. Strangely enough, I was comfortable with it.

SSG JJ squatted down next to me, so we were face-to-face. I wanted to rob him of his moment, slaughter his weakness, and destroy his arrogance and self-righteousness. While holding my cup, I leaned a little closer to make my words cut into him. "Fuck off, pussy, and stay in your lane."

He stood up, his face unreadable, and walked out. As I watched him go, exhaustion settled over me. The weight of it pressed on my shoulders, and I decided I'd had enough for the night. This town—this place—seemed to thrive on apathy, where only the strong and ruthless dared to impose order. *Fine*, I thought. *If they wanted their fragile, lawless bubble, I'll be the one to wreck it—and I'll do it loudly.*

I moved through the building, the tension a constant hum in the background. Upstairs, I checked on the positions, ensuring every angle was covered. The RTO confirmed that reports were being sent out, the steady rhythm of his radio chatter a reminder of the ever-present grind. The family huddled together, silent and wary. I gave the security team their final instructions, clipped and precise. By now, this routine was second nature.

I forced myself to stay awake until dawn, my body attuned to the eerie stillness of those early hours. When the first light crept across the horizon, I allowed myself to lie down, closing my eyes against the relentless fatigue. It felt like seconds had passed when I woke, roughly three hours later. My body ached with weariness, but there was no time to indulge it. I made my rounds.

SSG JJ greeted me with a look that could curdle milk, his expression grim, his demeanor radiating doom. Something had clearly set him off, but I brushed it aside. SGT TM held his position in one room, methodically going about his task. Inside the same room, SPC Bennett and I exchanged a nod, a quiet acknowledgment of the grind. In the adjacent room, SPC FG and SGT RB maintained their watch. Downstairs, SGT SH and the RTO managed the security element.

SGT SH was silent, his body language stiff with an edge of belligerence that was hard to miss. His mannerisms grated on me, a simmering annoyance I tucked away for later. I made a mental note to call him out when we got back. I didn't have patience for pettiness, not in this hellhole, not when every moment demanded more of us than we had to give.

The tension in the air was palpable, the kind that set every nerve on edge. It clung to us, unseen but suffocating, a constant reminder that we were deep in the belly of a place that would love to swallow us whole.

When I returned upstairs, I saw SSG JJ inside the room with SGT TOM and Bennett. SGT TOM was watching activity down the street. I decided to relieve him and give him a break. He quickly accepted the break, and all three of them were soon conversing.

For several hours, I watched through the opening of the curtain. Looking west about 300 meters, I observed an open market selling goods from meat to clothes. I watched until I got bored. I called out to Bennett and ordered him to take over. I handed over the laser range finder and provided some last-minute updates on the situation in the street.

"If that market empties out quickly, we are going to get lit up. So, keep a close eye on it and watch the surrounding area."

"Roger that."

I went downstairs to ask for updates.

"SSG Martinez, some people came to the door, so we took them in, and they are now with the family."

I walked over to the room and looked inside. The new people looked content.

I told SGT SH, "Any more people will probably compromise us. Get the terp to tell the homeowner to tell any other people to go away because several members of his family have a stomach virus and do not wish anyone else to get sick."

SGT SH didn't say a word, but I watched him obey orders. That was good enough for me. I went upstairs and checked on the second position. They appeared alert and in a good mood. I was then invited to lunch with the family. I was hungry. No one had tried to poison us, so I joined in.

We were seated at a large makeshift table. Soup and a plate of rice with meat were served. Like any good Arab, I ate it with my right hand, pinning my left hand nearly under my leg. The homeowner was so impressed with my Arab eating habits that he walked over, and we talked about life in good terms. I remember him clapping and enjoying his food. I smiled and thought how sad it was that we may have to kill his neighbors or destroy his house.

When I returned upstairs, I entered Bennett's position in the master bedroom. SGT TOM was sitting on the bed, and SSG JJ was nearby. Bennett was to my right, looking through a natural opening in the curtain. As I leaned back, my thoughts turned toward my upcoming leave—imagining a medium steak, nicely seasoned, with Spanish rice and a cold long neck.

The window approximately ten feet away exploded in a shower of dust, flash, and sparks. I immediately felt a stabbing pain just below my left knee, and the force violently drove me face down on the floor. My hearing was gone. Everything was in slow motion. In a few seconds, I heard the sound of shots fired, but they were from our side.

Damn, so close to leave, and my leg is fucked up. I rolled over and shouted, "I'm hit. I'm hit!" On my back, I looked to my right. Bennett was on his back with his feet on the bench seat, not moving.

SGT TOM was thrown into the wall, nearly upside down between the wall and the bed. He wasn't moving either. Through the dust and smoke, I saw SSG JJ. I called to him, "I'm hit."

SSG JJ was on his stomach, dragging himself using his elbows with his weapon in the crook of his arms.

"J!" I yelled, but I couldn't hear my own voice.

"Calm down. It doesn't hurt," I whispered to myself, trying to suppress the rising tide of panic within me. I focused on my breathing, inhaling deeply for a count of four, then exhaling slowly, allowing the rhythm to steady my racing heart. As I concentrated on the soothing breath, I watched SSG JJ struggle to drag himself out of the smoke-filled room, his movements cumbersome and labored. He didn't glance back, didn't check if we were all right. I felt forgotten, left behind in this murky, dust-laden space somewhere in Iraq. Abandoned.

Okay, fuck me. You have to look at it and do it yourself. Don't pass out, pussy.

I sat up and patted my leg. I found nothing. *Fuck. The wrong leg.* I checked my other leg. No injury. *What the fuck?* Grateful, I took a breath.

I rushed to SGT TOM, who had started to move. "You, okay?" I asked him.

"I'm still checking," he replied, more annoyed than concerned with his safety.

Two quick steps and I reached Bennett. He was sitting up. I was relieved. "You okay?"

"Yeah, what the fuck? I dropped my sandwich."

"Joey! Fuck the sandwich and check yourself!" I was pissed.

Since they knew where we were, I pulled the curtain down. The room lit up from the glow of the afternoon sun. I looked outside to my left and toward the market.

Huddled behind a rack of clothes was a man with an AK 47 shooting in our direction. I brought up my weapon, placed my red dot on his chest, and fired three slow shots. At the same time, I heard two more shots join mine. "Get some, FG."

The man fell back and did not move. I shot two more times, and his body fell between a rack of clothes. "You should have picked better cover, fucker."

"All right, Joey, get SGT RB and yourself and move to the roof. I will check on security downstairs. Then, I will come to join you. As I started to move downstairs, I heard someone yelling, "We are all good down here. No one is hurt!"

I called back, "Okay, we are good here. No one was hurt. I am moving to the roof."

I turned around to see Bennett and SGT RB stacking on the door. As it opened, I joined them, and we headed up the stairs. My head was on a swivel. I was looking everywhere to my right and to my left. Below, across the homes, no one was out. We made it to the roof without incident.

As we approached a wall, SGT RB called out, "We have movement with a MAM, a weapon, 12 o'clock, 150 meters behind a tree."

A series of shots.

Again, SGT RB spoke. He was eerily calm. "I got him. I think it was an RPG." He moved to the west wall and called out again, "Hey, large bongo truck with MAM's dismounting. It looks like they have weapons."

I ran over in time to see the last insurgents jump from the truck. Some ran to the far side of the street and some to the near side. I focused on a kid about seventeen years old. He was running toward the far-side alley, trying to find cover. He had on a white short-sleeve shirt, dark trousers, and sandals. His weapon was in his right hand, and his back was exposed to me.

I shot him. The first shot speared his back, and he stumbled. The second and third shots knocked him to the ground face down. He was still alive. He turned to look where the shooting was coming from, his face registering the confusion and fear of knowing he was dying. Somehow, he dragged himself to a set of boxes off the street. I never saw him again.

SGT RB was sending 40mm HE from his M203. I saw two explosions land in an alley where we saw a pocket of men running. As the dust settled, I heard painful cries and confused voices. SGT RB leaned against the wall and smiled at me. I smiled back. We were in control. I jumped on the radio and reported our situation to our security element on the first floor.

I saw Bennett loading his M203 with HE. He fired three rounds at the most western intersection. A truck driver had blocked the road with his dump truck.

"Bennett, kill that motherfucker!"

"I'm trying!" he yelled back.

A brown four-door sedan with two occupants tried to pass the truck at a high rate of speed. I ran to the west wall, leaned over, and shot the driver. The vehicle rolled to a stop. The front passenger leaned over the driver. Two occupants leaped from the vehicle. The driver was moving slowly, holding his hands near his chest and stomach. He was wearing a short-sleeved brown shirt and trousers with sandals. I shot him twice more. He leaned into a tree, dying.

I searched for the passenger, but he was gone. "Bennett, did you see the passenger who jumped out?"

"No."

I looked at SGT RB. He was absolutely cool. "SSG Martinez, you smoke?"

"No, brother, I'm good." I heard a click of a camera. Bennett had taken a photo of RB and me.

SGT SH stepped onto the roof headed casually to the wall. He appeared so casual that I got angry. "What the fuck are you doing? Aren't you in charge of the security element downstairs?

He ignored me.

Suddenly, a pigeon landed on the lip of the roof, looking straight at us. I peered over the wall and saw a man releasing his pigeons. I pulled out my pistol and tried shooting the pigeons, but I missed. The pigeon remained on the lip. Two more pistol shots, and the pigeon fell off the rooftop. The second apparently missed and ricocheted off the wall. I laughed at the situation. Chuckling, SGT RB holstered his weapon.

SGT TOM joined us. "We were shooting pigeons," I said.

He looked concerned. "Well, leave the fucking pigeons alone and let's check our sector."

I scanned our left flank and immediate front. No movement. An attack helicopter was making low runs over the neighborhood. The Iraqis hated the helos.

For the next forty minutes, we would take sporadic firing from different directions. Since we didn't have immediate targets, we held our fire.

As the helo continued to scan for targets, I decided to move off the roof. SGT JJ informed me we were being extracted by track vehicle.

We would link up with the vehicle two blocks away. The road in front was blocked by abandoned vehicles, and the road behind us could be full of IEDs. Moving on foot was the best option. We began to collapse systematically and pushed out the back door toward the street behind the yard. Since we expected a counterattack from our 12, I was one of the last to leave, along with SGT RB and SPC Bennett in a delay action to protect our 6.

As I prepared the guys to leave, SSG JJ was arguing with the homeowner. "Fuck you. I'm not giving you any money!"

As the conversation became heated, I asked the interpreter for the details. The homeowner wanted to be reimbursed for the damaged property. I agreed with SSG JJ, but we could at least pay for the meal.

I approached the homeowner. He suddenly became quiet and looked at me. I guess he was expecting another barrel under the chin. Considering the circumstances, I handed him almost all of my dinars because his house was heavily damaged. "*Shukran.*" Thank you.

SSG JJ took this as his cue and proceeded to move with the team. After ensuring personnel on the team were accounted for, I joined SGT RB and SPC Bennett, and we exited the property. I remember looking at the man's sons. Their eyes showed no emotion. The woman was crying, and the older man had his arms up in the air as though in prayer. I felt nothing, no anger. no regret, no joy. Nothing but pure cold numbness.

The ground before me was open, with a few palm trees between me and the next block of houses. I went into training mode. I heard myself, "SET, scan… MOVING... SET… scan... MOVING… SET… scan… MOVING…" This continued for two blocks. No one appeared tired. We were running with packs and a full combat load.

Eventually, I saw the M113. 1LT Vincent Caliguire, a New Jersey Soldier, a friend, and mentor, was manning the .50 caliber heavy machine gun. He looked deadly serious. He was doing his job.

I ran toward the vehicle. Passing Soldiers as the ramp dropped, I took my position outside, scanning the approaching team as they moved with practiced, controlled urgency. Each step was calculated—fast enough to get everyone on board, slow enough to avoid a deadly bottleneck around the vehicle. As the last man cleared the ramp, I followed, gripping the cool metal edge and hoisting myself into the metal beast.

Once inside, the vehicle lurched forward, engines roaring as we began the tense ride back to base. I steadied myself against the metal interior, every jolt a reminder of how exposed we still were out here, even with the armor around us. Quietly, I whispered a prayer under my breath, words that had become second nature out here. "Lord, get me home or make it quick. I do not want to burn to death."

This time, fate was merciful. We returned to base without incident. As the vehicle rolled to a stop and the ramp lowered, I felt the tension release from my shoulders. A weight lifted—if only for a moment. Another mission behind us, another day alive and victorious in Ramadi. This was all we could ask for.

After this mission, SSG JJ had a nervous breakdown and refused to lead his team. Consequently, he was transferred out of the Scout Platoon. I never saw him again.

Soon after that, I was out on a mission with another Soldier when we encountered an IED trigger man. When the device detonated prematurely, the man bolted. In that split second, both of us took aim. I can still picture the way the man stumbled, struck by our fire. I saw the fear in his eyes as he fell, still gripping the shovel he had been using to bury the device.

No one came to collect his body. He lay there in the broiling sun, untouched, a foreign fighter left to rot. In that culture, families would normally rush to recover their own, to make sure their burial was before sunset, a deeply rooted custom to honor and respect the dead.

When we returned to Camp Corregidor, Aguirre challenged our story of what had happened. Rather than face me directly, he went after a lower-grade Soldier. Aguirre's voice cut through the air like a bitter wind. He argued with the Soldier, his tone dripping with disdain. This was a thinly veiled jab at me, questioning the judgment that led to the deadly encounter.

I'd had enough. I told him he had no right to undermine a younger Soldier like that. "You weren't there," I snapped. "You weren't in the briefing, and if you ever do that again, we're going to have a man-to-man conversation, and it won't end well for you." To this day, I regret my words. I should have been more professional. He did not deserve my anger. If I wanted to be a leader, I needed to set the example.

However, I couldn't let his behavior continue. Hoping for a resolution, I took my concerns to his section leader, who told me that Aguirre would be transferred out of our unit to a conventional one, paving the way for a new leader to take over his team. That's not exactly the solution I was looking for, but my leave was coming up in a few days, and I had to get my team ready. Aguirre drifted to the back of my mind.

Leave

When my leave began on March 13, 2006, several days after the Diyala Zone 50-55 fight, I was lucky to be in one piece. I know this sounds callous, but I was not in the mood to head home. To me, that was another life. Shadow had been in the fight for nine months and had managed to walk away unscathed. It was only a matter of time before the dam would break.

Every step toward leaving felt like an act of betrayal. My mind told me I needed the break—my body was screaming for it—but my heart resisted, weighed down by the guilt of walking away while my brothers were still deep in the fight.

Rumors of what was coming spread like wildfire. Reinforcements were expected in April, and the siege of Ramadi loomed, promising to be even larger and bloodier than Fallujah. This was the crucible that would define us, the moment we would find out if we genuinely had what it took to endure the fire. The thought of not being there, of missing it, made my chest tighten, the kind of ache that lingered long after the decision was made.

Leaving the team while they were entrenched in Ramadi's chaos felt wrong on every level. I told myself I'd be back soon. They would carry on without me. Although physically removed, I remained emotionally shackled to every decision, every moment I missed. The guilt was crushing.

More than a battlefield, Ramadi was a part of us, its dust and despair ingrained in our skin, our lungs, and our minds. Leaving felt like abandoning something sacred. As I prepared to go, a part of me stayed tangled in the bonds of brotherhood and the jagged edges of conflict.

Still, I knew it was time for me to go home. Prior to leaving, I got the team together after training on the range. We sat under a swaying palm tree. I had to speak louder than usual to be heard over the wind.

I had given this some thought up to this point. Although SPC RT had been promoted to Sergeant, his performance had been substandard for the last few weeks. Considering the environment we operated in and the dread that something was going to happen, I decided to make some changes rather than follow the rules.

"Gents, listen, in less than 48 hours, I'm headed off to go on leave. SPC Bennett, you are in charge." I looked straight at him. "You are responsible for their health, welfare, and morale." I pointed to the other two. "Do you understand? Do you two understand? If Bennett says do something, act as if the order was mine. Bennett, when op orders come down the pipe, if at any time you think it's poorly planned, turn the mission down. I will back you no matter what. If you decide the mission is a go, ensure all coordination is done properly. Do not take any shortcuts. Delegate tasks and supervise. Do not do it all yourself."

Taking a breath, I went on, "We have been here since late June 2005, and here we are enjoying this sun in March 2006. Consider everything we have accomplished in Ramadi. We have reached many milestones as snipers. We have fought hard, and we have won every fight. But sooner or later, based on percentages, we may lose a fight. That's how the numbers roll. While I'm gone, be extra vigilant. Take care of each other and your equipment. The weather will warm up soon, and they will be itching for a fight.

"As always, be prepared mentally and physically. If the fight is not going well, you must force yourself to stay in the fight. Do you understand? *Stay in the fight.* Do not turn your back on them. Human nature will dictate the rest. Trust me, they will back off." I shifted my position and swallowed. "Okay, before I get emotional, is there anything I need to address before I leave?"

With nothing further to add, we got up and headed to the chow hall. I had done all I could.

As the weeks wore on, our reputation among the Screaming Eagles grew with every engagement, every calculated move in this deadly game. I often found myself in deep discussions with Gunslinger Company's commanding officer, a leader as charismatic as he was aggressive. He was the kind of officer who led from the front, inspiring his platoon leaders to match his tenacity. Gunslinger's area of operations was

Mula'ab. The 506 had committed two entire line companies to control it, a strategy that marked them as both bold and unrelenting.

I believed, and still do, that the key to controlling Ramadi lay in crushing the insurgency in Mula'ab. It was the heart of resistance, the nerve center of AQI's influence. When I was with 2/69 Mech Infantry, we never achieved the foothold needed to sway the populace, let alone instill order.

The 506 was different. They surged forward and caught the insurgency off guard. They struck hard and deep into Mula'ab, forcing the enemy to engage on our terms. The insurgents responded in kind, and the 506 suffered heavy casualties. Their resolve was tested in ways that most units would have balked at.

But they didn't break. If anything, the resistance only hardened their resolve. They made it clear to the people of Mula'ab that there would be no middle ground. It was a psychological war, a message etched into every operation: you are either with us, or you are against us. If an IED detonated beneath our convoy, we responded with overwhelming force—hitting from the south, the east, and the west. We came on foot, by air, and by vehicle. No avenue was unexplored.

Victory was the only acceptable outcome. Mula'ab would not remain a haven for AQI. It would not stay a stronghold of rebellion. The people would either adapt or find themselves caught in the crossfire. This sounds cruel and harsh, but war doesn't accommodate softness. The stakes were too high, and any sign of weakness would only embolden our enemy.

We had no illusions. Each patrol and each raid meant life and death. We walked into Mula'ab with our weapons at the ready, always prepared to react faster than the enemy. Our presence was deliberate and forceful. Dialogue was an option, but the outcome rested entirely with them. Fate lay not in our hands but in their decisions, actions, and silence.

Still, beneath the tactical clarity, a moral weight lingered. With every airstrike, every breached doorway, and every life taken, a question gnawed at the edges of our consciousness: Were we bringing peace, or were we sowing seeds of a deeper resentment? We had no room to entertain the question, but it lurked buried beneath the cold calculations of survival and mission success.

"Were we bringing peace, or were we sowing seeds of a deeper resentment?"

Mula'ab was a battle for control—over a city, over an insurgency, and over ourselves. And that battle could only have one victor.

The responsibility for what happened on Corregidor rested squarely with battalion leadership—and in this case, that was a good thing. Under LTC Ronald Clark, Battalion Commander of the 1st Battalion, 506th Infantry Regiment during the fight in Ramadi, leadership didn't just direct the fight. Leadership shaped it.

Clark and his staff created an environment where initiative wasn't punished. It was expected. They enabled, protected, and pushed the Recon Platoon to think critically and act aggressively, keeping the enemy—AQI—off balance, confused, and constantly under pressure.

That kind of command climate doesn't happen by accident. It comes from leaders who trust their men and understand that war is won by disciplined aggression, not hesitation.

Alongside him, Major Womack, the S3, held the operational reins with precision. Together, they turned intent into action and action into results. They didn't micromanage the fight. Instead, they empowered it, and because of that, success showed up again and again in violent, chaotic firefights where lesser leadership would have stalled.

There are accounts and analyses of Ramadi that highlight how adaptive leadership at the battalion level helped shift momentum during some of the war's hardest fighting. LTG Clark (He was later promoted to Lieutenant General) embodied exactly that—forward-leaning, engaged, and committed to winning at the point of contact.

From the ground, it was unmistakable.

To those under his command, there was no doubt—LTC Clark was in the fight. Not removed from it, not observing from a distance. He was fighting it, plain and simple. And for those who lived it, saw it, and fought through it, that kind of leadership was followed without hesitation.

Before leaving, I became consumed by the fraying edges of the team. Growing tensions had become impossible to ignore, especially between me, one of the team leaders, and the rest of us. SSG Cooke and I had spent hours discussing this Soldier's behavior—his cutting remarks, his thinly veiled contempt for the section, and his outright disrespect for SSG Cooke. This Soldier and I had been on a collision course for weeks, each interaction building toward something explosive. I knew it was only a matter of time, so I laid my concerns bare to SSG Cooke.

To me, SSG Cooke was a leader in every sense of the word. He was the steady hand that kept us from flying apart. Intelligent, decisive, and unshakably resolute, he could turn chaos into something manageable. Even in Ramadi's unrelenting violence, Cooke brought clarity and focus. To this day, I'd follow him through fire, no questions asked.

Not everyone felt the same. Some in the unit grumbled about his approach, but I saw nothing that justified their doubts. SSG Cooke was the anchor keeping us tethered. Without him, we'd drift into dangerous waters.

This team leader, on the other hand, was a thorn in my side. His most chilling outburst came just days before my departure. With an unnerving calm, he declared that if he were killed in Ramadi, heads would literally roll. The room went silent. His words hung in the air like smoke from a detonation. The cold, venomous conviction in his voice, the way he looked at us as though daring us to challenge him—I'd had enough.

When he walked out, my fury ignited. I took it personally because it was personal. He was threatening the people who fought beside him, the men who would risk their lives to bring him home. That betrayal stung more deeply than any insult because in combat arms, we were like a family. Such conduct burns deep because we love that Soldier. He is our brother.

One day, returning from chow and enroute to SSOC, I opened the topic with SSG Cooke. "What do you think of this team leader?"

SSG Cooke let out a sigh, struggling to say the right words. After a short conversation, I left and didn't speak of it again. Leave was on my mind. In addition, I was excited that I was feeling emotion, happy and relieved. Within a few short days, I would be home with my wife and kids.

Finally, the day arrived. I packed light and mounted the vehicle that would take me back to TQ. Once at TQ, we were housed in a transition barracks. I recognized no one, although we were all Screaming Eagles. The troop handler from the 506 appeared happy and content with his job. I could not help but wonder why he was not with us at Corregidor. We needed more Soldiers, and he appeared fit and organized.

As I tried to stay warm, I overheard his judgmental comments about a recent mission and that pissed me off. I blurted out, "Brother, you weren't even there. You're not tired. You're not cold, and you are pretty well isolated inside of TQ. I strongly suggest you choose your words with care." I did not want to fight another Soldier, but it needed to be said. He saw my rank and did not respond.

Next to me, a Soldier asked, "You're Shadow 4, aren't you?"

I looked up because I had not heard the call sign in a while. He was an LT, and he said, "We heard about you and your team. You guys are doing a great job." He introduced himself and told me he was headed on leave, going to Florida. He told me not to bother with the trooper. He was the local idiot, and no one cared about his remarks. He was here because he was worthless in the field.

The leave process involved nearly six days. Between leaving TQ, flying to Kuwait, and getting the "do not go home and kill your wife speech," it was a blur. I was operating on autopilot. I felt numb and had some anxiety. I was entering a new world that was supposed to feel familiar and safe, but I was different.

By the time I landed at Newark airport, I was alone. I barely remember my wife picking me up. I don't remember sleeping. Once I was home, I decided to visit my police department. As I was getting ready, I went downstairs to check my email. Standing at the desk, I clicked the icon and immediately saw an email from SPC JY.

Excited, I opened it… and lost my breath. Slowly, I sat down and read it again.

CHAPTER 11
CROSSING THE LINE

Shadow 4 Sniper engaging enemy target inside Mula'ab Jan 2005.

SSG Luis Aguirre and Joey Bennett were caught in an ambush. Aguirre was dead. Bennet had been shot, but the wound wasn't serious. My mouth went dry. My hands trembled. I could barely close the email.

If I had reported Aguirre sooner, he wouldn't have been in that place at that time. He would still be alive.

I tried to tell myself, *Nothing you can do*, but a wave of suffering swept across my soul. Anger enveloped me that day. Bracing myself, I slowed my breathing to bring my emotions under control. I still intended to keep my schedule for the day.

For an entire year, I carried that guilt. My slow, painful descent into grief felt almost unbearable. Despite our clashes, he deserved far more than what he received in the end. We were a family, and brothers fight all the time. Fifteen years later, as I write this, my eyes well with tears.

At nearly 10:00 that same morning, I arrived at the police station. Familiar faces welcomed me, trying to mask the weight we all carried. I forced an upbeat posture, but the effort felt hollow. When I encountered a detective I deeply respected—someone known for her unyielding compassion—a friendly exchange with her quickly shifted into shared sorrow. Grief swelled in my chest, threatening to spill out. Not wanting to break down in front of my colleagues, I ended the visit abruptly, retreating before the emotions overtook me.

I was shattered. None of us deserved to die in an ambush.

Until now, our losses had been men I only knew in passing. Each death hurt, but this was different. Aguirre was more than a name on a casualty list. He was my brother-in-arms. We had shared meals, traded stories, laughed, and argued like family. If I hadn't taken my irritation to our superior, if I had taken a different attitude toward him, if I had chosen better words—guilt-ridden thoughts crashed over me again and again. Aguirre's absence haunted me long after I left Ramadi.

Later that morning, I found myself parked in front of St. Paul's Church on Nassau Street, the engine idling as the rain drummed against the windshield. The droplets traced jagged paths, pooling into tiny rivers at the base of the glass. I sat there, motionless, staring at the church's steeple blurred through the rain. My breath stuck in my chest, heavy and sharp, until I finally exhaled a long, trembling sigh.

I prayed for Aguirre's family, for his children. I prayed they would somehow find the strength to bear the unbearable, to carry the weight of his absence that was crushing me. I thanked God for sparing Bennett, though the gratitude was tangled with guilt. How could I reconcile relief for one life saved while feeling grief for another lost?

This was war, its arithmetic cruel and indifferent. Survival was a game of relentless trade-offs, and the balance always felt skewed. Sitting there in the parking lot, I carried the weight of the entire war squarely on my shoulders.

"Survival was a game of relentless trade-offs, and the balance always felt skewed."

Memories came in fragments, sharp and disjointed, like shards of glass cutting through my mind—Aguirre's laugh during rare moments of levity, the way he carried himself during patrols, the sound of his voice over comms. He was gone, leaving a void no one could fill. My wife and family knew what had happened, but their understanding stopped at the edge of words. How could I explain the conflict inside me—the guilt, the loss, the gnawing sorrow because I had not done enough? How could I make them see the faces, hear the voices, feel the bond that tied us together in the chaos of Ramadi? I couldn't. I was alone.

I gripped the steering wheel, my knuckles white, and thought of my team—my section, my platoon, my family. That's what they had become to me. Family. I felt the pull to go back to them, to throw myself into the fight again, to be where I belong. I found no safety in this parking lot, no solace in these quiet streets. My place was with them, where the air crackled with tension, where the bonds forged in blood and fire felt unbreakable.

The rain continued to fall, a steady rhythm against the windshield. I sat there breathing, grieving, and bracing myself for the days ahead. The war wasn't over. Not for me. Not for any of us.

On my final night, my younger brother showed up. He and my son began to drink. As for me, I couldn't finish my third beer. I did something I never thought I could do. I poured it into the toilet and went to bed. I felt sick all night. Hours later, I threw up. I felt weak and confused at the state of health. It did not make sense.

I was still in the bathroom when my wife called my name. My Chief, CPT, and close friend LT Henderson all stood in my dining room. They came to say good-bye. I quickly brushed my teeth and rinsed my mouth.

I took my time going down the stairs. My brother was passed out on one couch, and my son lay on the smaller couch. The table was covered in empty bottles.

Embarrassed, I quickly apologized and tried to convince them that I had not drunk that much. Their grins told me I was wasting my time. After a few words of encouragement and comments to "get home safe," they left. I went back upstairs and resumed throwing up.

When my wife came in to check up on me, I felt ashamed, weak of spirit and heart. My body was refusing to accept my return to Ramadi. I said, "Babe, please get me to the hospital, I can't travel like this."

Eventually, we ended up in the emergency room at Princeton Medical Center where everyone recognized me. After all, I was a police officer for over ten years in that town. Eventually, I had to make the dreaded call and explain that I could not make my flight. They took my name and informed me to show up the following day if I felt better. One of the doctors said, "Officer Martinez, I know where you are headed. If you like, I can delay your return for three days."

I looked at him and quietly replied, "Give me some medication to control the vomiting. I intend to be on that plane tomorrow. Please."

The doctor nodded and walked away. On the TV up near the ceiling, I listened to a comedy act by Kathy Griffin and found myself chuckling. The medication helped, but that little bit of comedy helped as well.

My wife was also a major source of comfort. I needed her, and she was there for me. The next morning, she dropped me off at Newark Airport. I asked her not to come inside. She cried and pleaded with me not to get myself killed. She told me, "You bastard, you do everything you have to do to get back to your family. I don't care what you do, but make it happen." We hugged, and I walked away.

I was headed back to hell.

The following is the actual written statement submitted by Specialist Joseph Bennett regarding the events leading to SSG Aguirre's death.

I, Bennett Joseph James,
WANT TO MAKE THE FOLLOWING STATEMENT
UNDER OATH:

March 13th Operation Great White LB 4434898572

We exfilled OP Patricia at 2245 heading south along east side of an alley in zone 7. As we approached the intersection at the end of the alley, we made a halt. I saw the point man, SGT Row and 1LT Goshen's PEQ-2 IR beams sweep the intersection left to right. The IR beams crossed over a wall in the southeast corner of the intersection. It was 2 RPK's that opened up on us less than 100 meters away. All I saw was a wall of tracers light our formation up.

They were all around me. I remember feeling like I was knocked down from my kneeling position. The rounds were cracking and sizzling as they flew past my face. I could see them skipping across the road in front of me and they were bouncing off the wall behind me. I could smell them burning as they zipped all around me.

As soon as I recovered, I fired my M203 and a quick burst from my M16. All I could think about was getting off the street. I was terrified because all I could see were the traces that were ricocheting around me like a swarm of angry hornets.

I threw my weapon over the wall and jumped. That's when a round hit me in the right arm above the elbow. As I was going over all I remember was the burning sensation in my arm, and I remember SSG Irby getting hit and going down. When I got over the wall I tried to unlock the courtyard gate, but I couldn't, there were rounds bouncing and skipping everywhere. I could hear rounds impacting on the other side of the wall along with explosions on the street that I thought were grenades being thrown from the west side of the alley.

I then jumped a wall in the courtyard to the north. I got the courtyard door open, it already seemed unlocked and slid open. PFC Sanchez and SPC Brown started dragging PFC Reed and SSG Irby into the courtyard. We got the rest of his gear out of the street. PFC Sanchez started buddy aid on PFC Reed and SPC Brown gave buddy aid to SSG Irby. I gave an M4 to the interpreter and told him we really needed his help. We still kept receiving heavy small arms and grenades kept exploding in the alleyway. I went into the street and kept calling SGT Aguirre. Aguirre was my sniper partner, who was in front of me before we made contact and 1LT Goshen, but I couldn't see anyone, and no one answered.

I ran back into the courtyard because the enemy kept shooting down the alley wall. That's when someone yelled to close the courtyard gate. I knew that we had to get into the house because we couldn't defend ourselves and the wounded in the courtyard. I took my ICOM out of my ruck and established comms with 1LT Goshen and gave him a sitrep on our situation. He told me to keep fighting and that QRF was on their way.

I took out my IR strobe and I took PFC Sanchez's strobe and used them to mark the outside of our courtyard gate and put one in the street for QRF to find us.

SPC Brown and I broke into the house and quickly detained the family as hastily as possible. We got the front door open, and PFC Sanchez dragged PFC Reed inside and SPC Brown and myself dragged SSG Irby inside. PFC Sanchez, SPC Brown and myself got all the equipment out of the courtyard.

SPC Brown stayed downstairs with the wounded and PFC Sanchez and myself went to the roof to engage and provide security. We started to take multiple RPG fire from our west southwest. I called 1LT Goshen on the ICOM about what was going on with our assets and we kept engaging the enemy. I ran out of HEDP for my M203, and PFC Sanchez went black on ammo for the SAW. 1LT and the rest of the element were in a house to our south. We kept receiving small arms fire from southwest and across the street. Finally, our air came on station and were sparking the target house but were not firing.

I called LT Goshen, and he said they were going to drop an LMAV. The Cobra fired a couple of hellfire missiles and SPC Pushkin fired the SMAW-D on the target house. PFC Sanchez and I kept engaging the enemy with our M4 and M16. The Fixed wing aircraft dropped the LMAV on the target house and the Cobra and Huey gunship started gun runs on it as well. I remember seeing QRF and CASEVAC come off the Farouk and come to our position and that's when they took our casualties. PFC Sanchez and I remained on the roof and continued to engage the engage the enemy while CASEVAC took our wounded.

I heard tanks to our south and heard them engage with the main gun. I saw 1SG Klutts taking charge of the CASEVAC. I maintained comms with 1LT Goshen, and he kept saying we would get left behind and to sit tight till he said to leave. PFC Sanchez and I maintained our posture and kept engaging the enemy till 1LT told me to get the rest of our guys into the next group of vehicles for exfill.

We ran into the street and jumped into the LMTV and kept taking fire. PFC Sanchez and I kept firing out the back of the LMTV till the 240 gunner behind us started firing, then I remember us getting down and the rest of the soldiers got into the LMTV. I saw SGT Aguirre's pack and our M-24, but he wasn't there. I asked where he was. No one answered at first, then someone said he was dead. We exfilled back to Combat Outpost to the Battalion Aid Station.

Later, I learned that Bennett had taken six rounds. His plates (armor) had saved him. He was hospitalized for a wound in his arm, then returned to duty.

Over a year later, a phone call from SSG Cooke helped me cope with Aguirre's loss and accept the fact that his death was not my fault. For the longest time, I felt it should have been me.

Shortly after I left on leave, SSG Aguirre was transferred to a line unit. When the mission was generated, almost everyone had assignments, but Gunslinger Co. requested snipers. Since no one was available, SSG Aguirre was asked to participate in one more sniper mission, and Bennett volunteered to go along. If I had not gone on leave, I would have been on that street with Bennet instead of SSG Aguirre.

I Can't Kill a Man Like That

April brought a cold wind that bit through our gear. Rain came in fits, soaking the dirt until it turned to clinging mud. I longed for the familiar, punishing heat of the desert. The sharp chill seemed out of place, an unwelcome guest in the theater of war. One night, we gathered in the hangar, huddled together as someone decided we needed a few team photos. The moment felt strange—half camaraderie, half obligation—like we were trying to capture something fleeting, something the war wouldn't let us keep.

I was barely twenty-two hours back from leave and now I am boarding the back of a helicopter, the taste of home still fresh in my mind. We had no time to swap stories or talk about the different beers I had tried. The war moved on, and I moved with it, assigned to a mission before I unpacked—no warmup laps, no adjustment period. After checks, we clambered aboard the CH-46 helicopter, boots thudding on the ramp as we adjusted our gear and settled into the tight, uncomfortable seats. Eight of us were crammed into the space, the hum of the rotors overhead drowning out any attempts at conversation.

The flight was uneventful, the kind of monotonous ride that lets your mind wander enough to make you uneasy. Marine Corps pilots had casual confidence from countless hours in the air. Their calm did little to erase the tension that crept in as we approached Diyala. Our mission was straightforward: take out any IED teams operating in the area. Straightforward didn't mean simple.

I was in charge of a sniper team for the night. Alongside me was another E-6 scout tasked with leading us to the hide. SGT RH, one of the snipers, assured me the house we were heading to would meet our needs. We would find out soon enough.

The plan was grimly clever in its simplicity. At dawn, Infantry companies from the 506 would sweep through the neighborhood in force, boots hitting the ground hard and fast, drawing attention the only way aggressive patrols can. They were the bait, moving through Diyala like a storm, daring the insurgents to come out and fight. The catch, of course, was that this bait bites back—and hard.

As we neared the drop zone, the tension among us was an old companion. Casual moments in the hangar were behind us now, left somewhere in the dry humor of shared cigarettes and half-hearted jokes. Ahead lay the unknown. We moved toward it with the grim determination of those who had made the trip before.

The CH-46 touched down with a jolt, the rotors kicking up a storm of dirt and debris that disappeared into the darkness. I stepped off the ramp into the unknown, my NODs flickering to life, bathing the world in eerie green. Almost immediately, I realized the terrain was not what I had expected. Grass swallowed me whole, its blades brushing against my gear and muffling the sound of my steps. The sea of green obscured everything, reducing visibility to the faint flashing light on RH's helmet, bobbing ahead like a ghostly beacon.

We pressed forward, the uneven ground pulling at my boots as the team moved toward the road. When we reached it, I felt a brief surge of relief at something solid underfoot, but the tension never left. The lead elements peeled off into a palm grove ahead, their movements deliberate and nearly invisible in the darkness. The ten-meter spacing between us made it impossible to pinpoint the exact spot where they had slipped into the trees.

I stepped back a few meters, scanning the area again, trying to locate their entry point. My breathing slowed, the cold night air pressing me as I fought to keep calm. My cool, that thin veneer of control, was replaced by the gnawing realization of how exposed we were.

Somewhere out there danger waited, and we were moving closer with every step.

I was alone in a new area. I walked up and down the road, looking. Out of frustration, I started to whisper "Aguirre! Aguirre! Dammit, how did this happen? As I called out and searched, I started to plan my way out of here. Find a house, secure the family, kill anyone who tries to stop me, steal a car, and find my way to MSR Michigan. *Shit, why didn't I bring a map?*

To my great relief, RH stepped out of the palm grove. "Were you calling Silvia?"

I didn't bother to answer RH. We continued to our final destination as I hid my impulse to hug that bastard.

While the infantry moved cautiously through the neighborhood, we remained hidden, watching and waiting from our concealed positions, scanning for any sign of suspicious activity. The 506th Parachute Infantry Regiment (PIR) had earned a fierce reputation for their ability to provoke the insurgents into showing their hand. When AQI insurgents rose to the challenge, the 506 met them with unmovable ferocity, sending them to meet their Maker.

From its officers to its NCOs and every Soldier in its ranks, the message from the 506 was clear: Come out and fight. When they did, the battle ended on our terms. I thrived on that mindset. This was war, stripped of its complexities and politics, boiled down to survival and dominance. I didn't care about the Sunnis' resistance to our presence or their refusal to align with our way. Politics and civil affairs were someone else's concern.

"This was war, stripped of its complexities and politics, boiled down to survival and dominance."

My mission was simple and unyielding—to hunt and kill the enemy. Every action I took, every round I chambered, focused on that singular goal. In a city gripped by insurgent defiance, the 506 gave a simple choice: submission or destruction. We would outlast, outfight, and outlive anyone who dared oppose us. This was our war, and we were here to win it.

From the outside, the house appeared to be a two-story structure, but after rousing the family inside, we found it was a single-story building with only a roof above. We methodically cleared every room. We would allow no hidden threats or surprises that could jeopardize the mission. Satisfied with the security of the building, we established it as our base for the operation. During the night, the roof became our observation post, offering a clear view of the surrounding area. As dawn approached, we transitioned back inside, aiming to reduce our visibility and blend into the environment.

As I assessed our situation, a cautious thought surfaced in my mind—one that had been lingering since a long-ago conversation with SGT SH (Shadow 1) when he brought up an unconventional idea. The enemy expected American snipers to operate from elevated positions on rooftops or tall buildings. If we engaged targets from ground level instead, we could trick insurgents into thinking the shots were coming from one of their own, disrupting their response.

The idea had stuck with me, both for its ingenuity and practicality. I had discussed it previously with my platoon leader. While he hadn't explicitly approved it, he hadn't rejected it either—leaving the door open for me to exercise discretion.

Surveying our situation, I saw an opportunity to test this approach. The circumstances called for adaptability, and this tactic offered a chance to disrupt the enemy's assumptions. With that in mind, I made the call to implement SGT SH's theory.

This wasn't a decision made lightly. It was rooted in critical thinking—analyzing the enemy's patterns, leveraging the element of surprise, and trusting in the flexibility we had been trained to use in dynamic combat environments. This was war, and innovation was often the difference between success and failure.

As the day progressed and boredom set in, the team leader entered our hide. He was a staff sergeant with lots of combat experience and an excellent management style. He told me, "Harry, go upstairs. We have some guys crouching down in the field to our front. There is no reason for them to be hiding. Get up on the roof and see what is going on."

I considered his request. In broad daylight up on a roof, we had enormous potential to be compromised and possibly killed by some lucky enemy insurgent. However, as General Patton once said, "Audacity, audacity, audacity."

I grabbed my long gun, my rifle, and my spotting scope and proceeded upstairs. I slowly opened the door. With each inch, I listened to my exposed environment. As my confidence grew, I selected a final position and crawled to where I could observe the field. The world of insurgency is not as chaotic as it may appear to the untrained eye. It operates with structure, precision, and roles carefully crafted to sow confusion and strike when least expected. Each team within an insurgency serves a purpose, part of a deadly web designed to disrupt, destabilize, and kill.

The men in the field had already made their mistake. Their polished appearance couldn't hide the deadly intent written in their movements, 250 meters out. They weren't random laborers or innocent bystanders. They were auxiliaries, a critical cog in the insurgent machine.

What made auxiliaries so dangerous was their ability to operate in plain sight. Unlike armed combatants, they moved freely through villages, fields, and roads, blending into the civilian population. Those of us trained to observe always spotted indicators—the way their eyes moved, their body language, and, above all, their actions. It wasn't about what they carried. It was about how they carried themselves.

"It wasn't about what they carried. It was about how they carried themselves."

They didn't openly hold rifles or wear uniforms. Their battlefield skillset was subtlety and deception, using everyday actions to serve their cause. They planted IEDs under the cover of darkness, delivered messages between cells, and ferried weapons to fighters waiting for the right moment to strike. They were the invisible hand of insurgency, supplying the combatants without drawing attention to themselves.

The team in the field had textbook auxiliary behavior. The lookout, sharp-eyed and vigilant, scanned for any sign of American forces while the others worked. The men digging in the dirt weren't planting crops. They were likely uncovering explosives designed to rip through convoys and claim lives. Their sharp dress, neatly combed hair, and polished demeanor were part of their calculated attempt to blend in, but their movements betrayed their purpose—the urgency in their work, the coordination between them, and the tension in their bodies.

Insurgencies thrived on the strength of their teams, each piece interdependent on the other. Neutralizing one role—whether a fighter, auxiliary, or leader—sent a message that the darkness isn't safe. As I watched the men in that field, their role became clear. Even clearer was what needed to be done. The battle against insurgency meant seeing the unseen and stopping them before they could strike.

The field offered no cover. The sun exposed every movement in stark detail. Two of the men stood as sentries, their backs to my position, while the third knelt low, digging with an intensity that spoke volumes. The crouching man finally rose, his companions moving to help him lift a heavy object wrapped in a something resembling a brown wicker basket with heavy woven texture and a carrying handle. The weight made them lean awkwardly, their movements laborious.

My eyes narrowed, my instincts sharpening. Something was off. The sun's angle revealed faint shadows beneath the bag—long, cylindrical shapes with a faint green hue. My mind raced. Farming tools? Equipment? But my doubts were quickly overshadowed by their movements. I could not ignore the way they moved, the calculated coordination of their actions.

Then, it happened. One of the men shifted, tilting his handmade carrier just enough for me to glimpse what lay inside. Suddenly, I gasped. Protruding from the edge of a wicker basket was a cone-shaped object, unmistakable in its design. A rocket. My focus narrowed like a vise, and all hesitation vanished.

During this observation, the security team leader (Staff Sergeant) and RH joined me on the roof. RH was to my left. In the softest of tones, I told him, "Left to right, 1,2,3. Kill number one, and we work our way to the middle."

At that point, the patrol leader made it clear he wanted to see for himself, and we ended up bickering, sharing an exchange of low-tone angry words and facial expressions about whether to shoot. It ended when he said, "I am in charge!"

He grabbed the binos. I guided him to target one. Finally, the patrol leader agreed that the ROE had been met and that we could shoot, but by now, the guys were moving away. The window of opportunity was closing.

I told RH, "Kill number one." RH moved to a kneeling position. He was having difficulty with the shot. Suddenly, he stood up to a low crouch, moved to my right side, and repeated the process. I was getting frustrated with both of them, but I did not want to instigate an argument. I remained silent and kept my position.

RH looked at me and said, "I can't take the shot."

I did not ask why. They were getting away. The enemy was still oblivious to the fact that they had been compromised. I didn't ask permission. In one motion, I grabbed my M16A4 and got into a hasty kneeling firing position. The men were in tall grass with their upper torsos exposed.

I didn't wait for confirmation. We had no room for doubt anymore. The certainty of what they carried burned in my mind, every detail locking into place. I put the red dot of the Aimpoint optic on the chest of the man with the rocket. He was number one. My sights settled, my finger found the trigger, and the world around me blurred into nothingness.

They were walking at right angles to us. I added a little lead with the edge of the red dot just in front of his chest. The round travels over 3500 feet per second. At a distance of nearly three hundred meters, the round would impact in less than a second.

I rested the handguards on the wall and supported the butt of the rifle with my left hand. I leaned into the weapon, let out my breath, and paused to reduce the red dot movement. The intended target's chest touched the left edge of the red circle. I squeezed the trigger and held it back inside the trigger well, allowing harmony in the form of death to properly deliver its message.

The body of number one man reacted—I hit him—so I shifted to number two. He made it easier because he turned around and leaned over to the fallen man to assist him. I squeezed the trigger a second time. He grabbed the fallen man, and all three disappeared into the palm grove.

We were using 62-grain projectile 5.56mm ammunition for our M16s and M4s. Those bullets would go right through people who would sometimes not immediately know they had been hit. I knew I had hit number one, and I felt I had hit number two, but in either case, they all disappeared, using the terrain to conceal their movement.

Combat tactics and leadership are intricately tied to simplicity, communication, and the seamless integration of roles, especially in high-stakes situations where seconds can mean the difference between success and failure—life and death. These events revealed a breakdown in operational control, one that underscores the importance of defined responsibilities and decisive leadership within a combat unit.

After the engagement, I reflected on what went wrong: the lack of a clear chain of command between scouts and snipers. While my natural inclination is to build trust and allow roles to flow naturally, combat has no room for ambiguity. On a mission, leadership must be explicitly established, and operational roles clearly defined, especially when multiple specialized teams are involved. Trust is vital, but it must be balanced with structure, ensuring that every individual knows their responsibilities and who they answer to at every stage of the operation.

"Trust is vital, but it must be balanced with structure."

In combat, the chain of command is a safeguard. If a leader is killed or incapacitated, the next highest-ranking individual must take over seamlessly. This principle of succession ensures that even under the most chaotic conditions, someone is always in control. When roles overlap or are poorly defined, confusion can erode the efficiency of a mission. That's exactly what happened here.

To correct this, we established a clear operational framework. The scouts, with their expertise in movement and reconnaissance, would retain control during ingress and egress, ensuring we reached the hide and exfiltrated safely. Once inside the hide, tactical operations would shift to sniper control. Snipers are specialists in observation and engagement. Their precision work demands autonomy during critical moments. Once the mission was complete and we left the hide, operational control would revert to the scouts. This division of responsibilities allowed each team to operate within its expertise while maintaining an overarching chain of command.

Combat leadership is about preemptively eliminating the potential for confusion. When roles and responsibilities are unclear, chaos creeps in, and chaos on the battlefield can be fatal for a team. Leadership means not only trusting your people but also providing them with the structure to operate effectively. It's a delicate balance, but one that's critical to the success of any mission.

"Combat leadership is about preemptively eliminating the potential for confusion."

I was slightly annoyed with that entire situation. I stared at both Soldiers, but I said nothing. A report with a description of the MAMS was radioed to our higher-ups, and a situation report followed.

After I cooled down a bit, I went back to the roof and spoke to the security team leader. "Look, I really like collaborating with you and hope to do so often in the future, but you have to respect my judgment when I tell you we are about to kill someone."

"You're right, Martinez," he admitted, his voice heavy with self-doubt. "I don't know why I did that. For some reason, I couldn't bring myself to shoot that man." His honesty was disarming, and I found myself admiring his integrity. His willingness to articulate something so deeply personal was a rare kind of bravery, to face both the enemy and the conflict within himself.

His words reminded me of stories I had heard before—of men unable to kill unless directly provoked, their instincts tied to self-defense rather than offense. This wasn't weakness. It was a deeply human response to the thin line we walk as Soldiers and the moral weight of taking a life.

That's when I realized I had crossed that line without noticing. In this hell called Ramadi, I no longer needed to feel threatened to pull the trigger. The transition was imperceptible but absolute. I was no longer a police officer trying to protect and save lives. I was a sniper—a disciplined rifleman operating with precision under pressure, focused on identifying threats, a hunter of the *homo sapiens,* killer of human beings, a dangerous beast on a death ground, destroying lives. My focus was on survival, necessity, and discipline to win this fight and the next fight. This was the reality of the war we fought, and I accepted my place in it, not as something to glorify, but as something to carry.

CHAPTER 12

WALKING WITH GIANTS

HHC RECON PLT assembled for a raid, Ramadi, Iraq, April 2005

Later that day, a report came through—a single male was seen carrying a body that had clearly succumbed to an open chest wound. A third individual, alive but severely injured, was reported with an open fracture on the right arm and a gunshot wound. When asked for a description of their clothing, we relayed the details, and they matched the trio we had engaged earlier. However, no bags or devices were recovered from the scene.

We had the sobering confirmation of our actions—a reminder of the split-second decisions we were forced to make in the uncertainty of war. The absence of the devices didn't negate the threat we had identified, but it underscored the gray areas in which we operated. In Ramadi, clarity was rare, but the stakes left no room for hesitation. We acted on what we saw, and the outcome was final.

Entry Control Point 8

After suffering a series of events ranging from triumph to tragedy, we established a new Entry Control Point (ECP) at a four-way intersection. The morning had lost its early chill, and heat rose with the sun. From our conversations, many of us were feeling the nearness of home.

During the last phase of Joker 4's sniper operations, I still lived with Dealer Company. The 506 D Company NCOs were at another tier level, building cohesiveness and brotherhood while maintaining lethality. Dealer Company was a machine, relentless and methodical, grinding through the days and nights with grim determination that bordered on inevitability. For all their grit and efficiency, their humanity shone through in the quiet moments—SGT Behrends's relentless humor, and Swadley, wise beyond his years. Where did they find these giants?

Although I often thought of moving in with the 506 scouts and snipers. I loved watching Death Dealer Company work. To this day, my reverence remains within the realm of awe and respect. Sure, sometimes we had disagreements and different perceptions about the war here in Ramadi, but generally, we got along well. They treated my team with respect and love, and I greatly appreciated the camaraderie.

Our AO had become more dangerous, and our missions were stacking casualties on both sides. Yet, the 506 refused to yield any ground. They drove on with the mission. The operation tempo was high, all strategic force multipliers were still on station—Seal Team 3, Air Force TACPs, and Task Force 145. Task Force Dark Eagle was still operating. Members of the original team members had been replaced. Unfortunately, the ANGLICO Marines were never replaced.

The 506 snipers had recently moved out of FMJ building to HQ building. Inside their new space, they built two floors and maintained an open floor. This is where the teams would position and display gear, allowing PCIs/PCCs to be executed by squad leaders and team leaders. Adjacent, facing the west wall, was the Scout/Sniper Operations Center. Any time I walked in, I always saw activity. LT AJ had taken us to the next level. Many names stand out in my head. I'd feel remiss as a leader not to mention these men.

May 2006 arrived. We knew enemy activity would increase inside Ramadi. ECP 5 had collapsed, and with that decision, ECP 8 would be stood up. Joker teams were ordered to provide overwatch for the assigned Infantry company.

An entire company was devoted to the task, and SSG Cooke assigned me an overwatch position north of ECP 8 for twenty-four hours. I considered the mission low-key. The proximity of an entire infantry company nearby would lower any concerns the team might feel. The death of Aguirre, the nervous breakdown of another Staff Sergeant, the injury of SGT Hughes (a Soldier's Soldier), and another Soldier on convalescent leave—all had impacted the entire team. Unfortunately, it would not end there.

Of the new Soldiers selected to join our scout and sniper PLT, one Soldier stood out. A young kid, PFC WD was a former West Pointer, dismissed from the school for misconduct (underage drinking). The Army decided to strip away his privilege of attending West Point and punish him. They assigned him to the 506, and here he was in Ramadi.

WD was always smiling and open about his service despite the setbacks. I admired that attitude, and I needed positive energy from him. When he asked to be one of my scouts on security, I initially felt reluctant. He was a new guy with virtually no experience here and full of potentially romantic ideas about war. I did not want to carry that burden with me. So, I said, "I will think about it."

As WD walked away, I yelled, "Hey, come back."

He quickly returned.

I told him, "WD, I don't know you. I do not know your skill level, your skill set, your mindset, or whether you comprehend what you are really asking here. Out there, people are trying hard to kill you. Well... all of us. When those rounds come at me, it feels personal. Like the enemy out there is focused on me. Stay with your team. In the meantime, I will watch you and speak to your leaders about your performance. If I like what I hear and what I see from you, I will ask you to help us out. Fair enough?"

WD maintained his military bearing. He had gone to the parade rest position and replied, "Yes, Sergeant."

Taking his cue and professionalism. I stared at him and answered, "Dismissed."

He moved to the position of attention, snapped an about-face, and walked away. I put my head down and chuckled to myself. Fucking West Point.

SGT J Hughes was charismatic and quick to smile. Tall with blond hair and piercing blue eyes, he was born with a chew in his mouth. He had the respect of the men and his leaders. He was audacious and open-minded, not to mention considerate. From time to time, I would join their team at the chow hall.

The bond of brotherhood in the infantry is crafted through the fires of purpose that unites us all—facing danger side by side and building unwavering trust. Each moment together strengthens these ties, creating a unique camaraderie that stands resilient against any challenge. In Ramadi where death was a constant shadow, this bond became our anchor—a connection that transcended race, religion, and creed. Out here, none of that mattered. We wore the same uniform, bore the same flag on our shoulders, and shared the same fight. Brotherhood wasn't just camaraderie. It was a way of life.

"Brotherhood wasn't just camaraderie.
It was a way of life."

On one occasion, a group of us gathered around a table toward the back of the room, close to an exit. The air was thick with the low murmur of voices, the conversation revolving around tactics and how to refine them. These discussions were as much about preparation as they were about staying alive. At the center of it all was SGT Hughes, his smooth southern drawl cutting through the room with calm authority. Hughes had a way of facilitating these talks, leading without dominating. His voice never rose in arrogance. Instead, he carried a steady confidence that made everyone feel grounded. He was a leader of warriors.

We sat hunched over the table, instinctively leaning closer as we spoke in muted tones. Operations Security (OPSEC) was second nature by now, even in the relative safety of this room. Vigilance was our default setting. At times, the heaviness of our conversations gave way to moments of levity. This was the infantry way. Grunts would be grunts, and even the most serious plans could be punctuated with sharp humor.

Someone would say, "Why can't you be normal?" or "I'll need your sister's number," followed by a low ripple of laughter. It wasn't disrespect. It was how we kept ourselves human in a world designed to strip away our humanity. It was how we found moments of light in the darkness.

In this makeshift room, we were equals, not by rank but by the unspoken understanding of what it meant to be in the fight together without the lines drawn by the outside world. Our backgrounds, beliefs, and differences were irrelevant. All that mattered was the brother to your left and right, and the flag on your sleeve. In Ramadi, we didn't just wear green. We lived it, embodied it, and carried it together, one mission at a time. Each of us was a crucial part of the whole, each of us valued for our unique contribution.

"All that mattered was the brother to your left and right,
and the flag on your sleeve."

This was the brotherhood of the infantry—raw, unfiltered, and unwavering. It was built on trust, sharpened by adversity, and reinforced every day by the knowledge that, in the fight ahead, we would have no one else to rely on but each other. That was enough. It had to be.

One day, I had my back to the wall when an older Soldier entered the chow hall and scanned the room. He spoke to a nearby Soldier. The second Soldier scanned the room and pointed in our direction. When the older Soldier thanked him, I realized he was looking for us.

When he reached our table, I recognized the rank of Command Sergeant Major (CSM). I announced his presence, and the snipers came to their feet. He motioned us to remain seated and introduced himself. "I am from Multi-National 5th Corp Iraq. I am looking for SGT Hughes." Immediately, the scrape of a chair sounded, and SGT Hughes stood.

"Sit down, SGT Hughes. I spoke to your leaders, and your name was brought to my attention many times regarding your professionalism and contributions as a Soldier and a leader. I want to show my gratitude on behalf of 5th Corp and present you with the CSM coin." It was the size of his palm, in the shape of a blood red Roman Numeral V. On the flip side was the CSM's rank and the words "For Excellence And Achievement."

The CSM extended his hand. As they shook hands, the coin passed from one palm to the other.

SGT Hughes said, "CSM, thank you. I appreciate your words and those of my leaders. However, if anyone deserves this coin, it is SSG Martinez. He has been here longer than any of us. He led numerous sniper missions, producing a significant death toll amongst the enemy. I cannot accept this. With your permission, I would like to present it to SSG Martinez."

Words left me. The Soldiers at the table were smiling and nodding. I stood up and looked at SGT Hughes. "Brother, you earned it. It's yours."

Holding it out, SGT Hughes said, "No. Take it, Goddammit!"

We all laughed. Awestruck, I accepted it—not because of the CSM but because of SGT Hughes (promoted later to Command Sergeant Major). Thank you, CSM Hughes. To this day, that is one of my proudest moments. Recognition from peers surpasses any award the system can give.

> "Recognition from peers surpasses any award the system can give."

SGT TOM is another warrior. He is intelligent, soft-spoken, quick with remarks, and uncanny with his sarcastic abilities. During one operation, we were on a rooftop in Mula'ab observing suspicious activity while an American patrol delivered medical bandages to a local clinic. As the company CDR was outside the clinic, they came under attack.

I watched through my binos as CPT OS a tall bull of a man, no-nonsense and straightforward, remained outside under enemy fire. He casually opened the driver-side door and initiated a SITREP (Situation Report). I saw his hand suddenly snap in the air. Something had struck him. He reached into the vehicle and came out with a bandage and wrapped his hand.

I stepped back and whispered my observations to SGT TOM. He replied, "We are spotting enemy observers around our position."

"Show me," I said.

At the southeast corner, he said, "Look out about 150 meters to the left. You will see a kid peering around the corner. He is wearing a dark-colored waist-high jacket, black pants, and black shoes. He appears to be ten or twelve years old.

I spotted him. "I can't shoot him. He is unarmed. Maybe he is just being stupid."

"I don't want you to kill him. I could do that myself. Look closely at his neck."

I adjusted the optic to 40X, my breath steady as I scanned the intersection. The boy was leaning against the wall, his small frame awkwardly positioned at the edge where concrete met air. Something felt off. As I focused in, a hand appeared—gripping the collar of his shirt, holding him firmly against the corner. It seemed deliberate and controlled. My senses came alive. I needed my rifle.

The hand vanished, replaced by an arm that snaked around the boy's neck in one fluid motion. An adult figure stepped into view, partially shielding himself behind the child. He was lean, his short curly hair and faint mustache matched the hundreds of faces I had seen in this city. But this man didn't blend in now. He was methodical, calculating, cruel, and a coward. I reported my observations in a soft monotone.

Through the scope, I refocused on the boy. His face told a story more vivid than any radio report. Fear was written in the wide circles of his eyes and the quiver of his lips as they moved in a frantic rhythm. I couldn't hear him, but I didn't need to. He was pleading, maybe praying. His shoulders were hunched up, his neck swallowed by the tension of his small body, trying to make himself even smaller.

The adult behind him was controlled, precise, and cautious. He was orchestrating the scene, leveraging the optics of terror. We had seen this old trick before but never grew numb to it. My hands remained steady, my breathing even. Training demanded clarity, and clarity demanded detachment. The boy's terror had to remain a detail, a data point in the equation of my decision-making.

Through the glass of the scope, I locked eyes with the boy. He could not see me, but I felt his gaze all the same. My mind raced ahead, anticipating the next move, calculating trajectories, collateral, and outcomes. This was no longer about the boy or the man holding him. It was about the hundred lives they threatened to disrupt if they carried out their intentions, a message as deadly as a bullet.

My breathing was heavy, the kind of rhythmic pounding that came from more than physical effort. I forced myself to slow down, inhaling deeply, then exhaling in measured control. Everything was deliberate now—every movement purposeful, every thought razor-sharp. I needed the best vantagepoint. I reached for my M24, cradled the weapon like an extension of myself, and grabbed my assault pack. The pack landed on the ledge with a gentle thud. I rested the rifle on it, ensuring the barrel remained free of obstruction. I steadied the rifle and prepared for the shot.

Through the scope, the scene came into sharp relief. The man's grip had changed. His arm now coiled tightly around the boy's neck, his other hand pressed firmly against the child's waist. He was leaning the boy forward, like a makeshift shield. His jacket sleeve slid up, exposing his hairy forearm that caught the sunlight, a cruel detail that seared the moment into my memory. The man's head hovered just above the boy's right shoulder.

My mind snapped to training, to repetition, to muscle memory. *You can do this.* I told myself, as much an order as a reassurance. *You've trained for this. You've prepared for this.* Police sniper scenarios flashed in my mind, the clean confines of a range replaced by the dusty chaos of Ramadi. No wind. Think triangle—eyes, nose, mouth. My finger hovered over the trigger guard as I made my final adjustments.

I toggled the cam to one, ensuring the elevation matched the distance. My windage was zeroed, no value. The math was perfect. The conditions were ideal. I mentally rehearsed the steps: Breathe. Aim. Relax. Squeeze. The mantra played on a loop in my head, a lifeline in the swirling storm of emotion and instinct.

Through the scope, the boy's terror was so vivid it almost hurt to look. His lips were moving, trembling as he mouthed silent words. His wide eyes darted around, searching for something, someone, anyone. They wouldn't find me. I wasn't a savior in this moment. I wasn't a person at all. I was the silent watcher, the hand of certainty.

The man's movements were subtle but deliberate, shifting his weight, calculating angles just as I was. My finger slid to the trigger. The rifle felt heavy. Its warm stock pressed my cheek and shoulder as if daring me to hesitate, but hesitation wasn't an option. Not here. Not now. The child's life was a pawn in the game the man was playing. The boy's survival rested on my ability to make this one perfect shot.

I exhaled, letting the tension drain from my body like a shadow retreating with the sun. My crosshairs steadied, aligning the triangle with precision. The world narrowed to a single point—the space between the man's head and the boy's shoulder. Everything outside the scope faded.

I steadied my grip, exhaled again, and squeezed. The man dropped to the sidewalk. The boy's hair appeared near the bottom of my view. The boy was alive. I shifted my focus back to the adult, tracking his every move. His feet twitched, awaiting death.

The kid ran away.

With my heartbeat thumping in my ears, I slid down the wall and looked at SGT TOM. In a flat voice, I said, "Hit."

We were close, about 140 meters, and I had stayed up too long. I knew someone had spotted me. *Prepare to defend our positions.*

"Roger, that was fucking cool." SGT TOM moved away and verbally reported my actions. Our Soldiers enhanced their fighting positions. Almost immediately, we saw movement. Iraqi males were in the alley, backs against the wall, creeping toward us. One was within hand grenade range.

SGT TOM was watching me. I grabbed a hand grenade off my vest, flipped it upside down in my left hand, pushed the clip off, and pulled the pin. I went to a kneeling position, then stood up quickly. I saw him creeping along the wall, looking up at the roofs around him. I went back to kneeling and threw the grenade into the alley. It landed in mud and water between the two walls and the man.

The grenade never exploded. I was dumbfounded. SSG TOM moved close to me and spoke. "Did you forget to pull the clip?"

"No. Fuck my life, I think I forgot to pull the tape!"

Then, the shooting began, and we responded in kind. We were able to repel any further assault. Soon, we decided to collapse our position and requested extraction. As we gathered our equipment, SGT TOM approached me. He said, "The shot was very cool. The hand grenade freaked me out! The grenade *not* exploding—very uncool. You need to work on that part." That was all he said, and that was enough for me.

A Wall of Flesh

During the construction of ECP 8, the sharp clang of hammers and the grinding hum of generators filled the uneasy silence. Dust swirled in the air, clinging to sweat-streaked faces, blurring the already fading boundary between man, machine, and sandbags. We all wore the same haunted expression, a mask forged in the furnace of unrelenting stress. Conversations were sparse. Words felt too fragile to bridge the void between survival and despair.

During my rotations at ECP 8, I scanned the streets that felt like veins pulsing with unseen danger. Ramadi was alive, breathing in shallow, jagged gasps as insurgents slipped through alleys and markets. The ache of loss was a constant undercurrent, a dull throb in my chest that threatened to swell with every new radio call. But I couldn't let it out. I wouldn't. Emotions were luxuries for someone else. Someone safe. Somewhere far from here.

We were trained to endure, to press forward no matter how hard our spirits became. But as the days wore on, I wondered if endurance was a polite word for surrender. The hope of something beyond this fight had faded. In its place was cold, sharp resolve. We would kill them before they killed us. Nothing more, nothing less.

During ECP 8 operations, the structure was built at night to reduce casualties. During the day, Soldiers would continue to build up the ECP, implement a rest plan, and prepare for the pending attack.

AQI attacks came like clockwork, timed to exploit our vulnerabilities. RPGs arced down the street in slow, deadly spirals. PKM bursts and AK-47 fire echoed off the concrete like the percussion section of a grim symphony. They meant each assault to wear us down and test our lines.

Sniper teams were deployed in greater numbers during the construction phase, the weight of overwatch falling heavily on our shoulders. Joker 4 was assigned the northernmost sector of the street—a critical position. The house we selected had commanding sightlines, offering almost a full 360-degree view of the surrounding area, the kind of location snipers dream about but dread to occupy. The rooftop walls were dangerously low, with no room for standing or crouching. Every movement had to be calculated. Every second, we stayed prone to minimize our exposure. As morning turned into afternoon, the rooftop radiated the heat of the blazing sun until we felt like we were lying on a furnace.

As I scanned the streets through the scope, my focus sharpened on a group of children moving through the intersection. They stood out immediately. They were uniformed, organized, and oddly clean amidst the dust and ruin. Each child wore white shirts with blue tops, shorts, and almost new backpacks. Their presence was both harmless and unsettling, a striking contrast to the chaos that often spilled into these streets.

The details came into focus, one by one. Five men walked amid the children like shepherds among a flock. Light-colored *shemaghs* wrapped loosely around their heads, the subtle movement of fabric catching in the slight breeze. They had white beards and lined, weathered faces and wore traditional garments in shades of white and tan, blending seamlessly into the dusty hues of the environment. But it was not their appearance that caught my attention—it was the way they moved.

I pulled out my map, noting their location and calculating the distance: 825 meters. They were on the edge of effective range, a challenging but feasible shot if it came to that. I adjusted the optic and scrutinized their movements further, looking for the tells that would confirm or dispel my suspicions.

The children provided perfect cover, an organic shield that would make the most seasoned sniper hesitate. This was deliberate. It had to be. Again and again, we saw this AQI strategy of leveraging innocence as a weapon. For me, that burned in a way no amount of training could entirely erase.

I wasn't alone on the rooftop. SGT RT and SPC JY were on my left, their rifles trained on overlapping sectors of fire. On the second floor below us, three Soldiers from Animal Company stood watch over the courtyard, maintaining radio contact and security. The embedded journalist—a rare breed of courage or recklessness—had accompanied us on this mission. He sat quietly, watching us work but never speaking. Most avoided missions like this, but not him.

Sweat trickled down my back, soaking into my gear as I shifted slightly to relieve the pressure on my elbows. The men and children continued their slow progress, each step a study in calm deliberation. Through the scope, I tracked their every movement, searching for anything out of place. At that distance, I strained to determine bulges that could conceal weapons and gestures that betrayed intent. My mind worked in overdrive, cataloging and analyzing every detail, knowing that one wrong assumption could cost lives.

Each second dragged, I watched and waited. The 40X power optic felt heavy despite being steady and light with my touch. The sun climbed higher, casting long shadows across the rooftops. My instinct convinced me to move toward my rifle. Immediately, I found my position. My trigger finger rested lightly against the guard, poised but patient.

Most people never understood this part of the job—the waiting, the watching, the remorseless cold precision required to sift truth from deception and danger from distraction. They looked far away when viewed through that 10x rifle optic. I placed my rifle nearby and grabbed the handset on the radio. I had questions, and I needed answers.

I called SGT RT and asked him to confirm the range and report our observations to Currahee X-Ray, BN TOC. SGT RT confirmed the range.

I told the guys nearby, "Damn, they are far away. Is there a school in that area? I know the river is close by. Why are there so many older males with these kids?"

Currahee reported back, "Continue to monitor," and added, "Action is occurring within Eagle (Marine) country."

"Roger, can you check if they have any friendly patrols near the area?"

After a fleeting time, BN came back and reported, "Negative, no friendlies nearby."

Continuing the conversation with the BN TOC I said, "Roger, I hate to bother you, but can you ask if there is a school near their vicinity? Stand by for the grid." I provided the 8-digit grid, added our position, and delivered a 10-digit grid.

The BN TOC asked me to define "vicinity." They needed me to paint a picture for them. I also wanted to erase doubt from my mind.

"Currahee X-Ray, determine if a school is nearby within two klicks of the stated grid."

I heard SGT RT's voice, "They stopped on the road."

I put the handset down and ordered SGT RT to monitor the radio. I picked up the 40-power optics and locked it into a low-set tripod. I started to glass the large group, moving in and out of the different magnification powers. They walked on what appeared to be an unimproved road. On the far side of the road was a palm grove. Adjacent to the road were wooden poles holding electrical lines. The group stopped near an electrical pole.

"BN reports they are unaware of any school within two klicks." That was over a mile, a long walk for such small children.

Glancing at SGT RT, I responded, "Roger, thank you."

The reporter was now sitting with legs crossed. He asked me what I was thinking.

I felt a little annoyed, but I answered him. "Why would little kids be out in a school uniform with backpacks so far away from school when they should be headed home?"

"Hmm, answering a question with a question," he shot back. Now, he sounded annoyed.

I replied, "You asked me what I was thinking, and that's it. I am attempting to apply a little critical thinking here."

I resumed glassing my area of interest and watched as some kids had their backs to us while others had their attention drawn to something on the ground. Two of the men had disappeared. Crouching, one came back into view. He turned a kid around, and there appeared to be some furtive movement. Then, he disappeared.

Again, I checked my map. My calculations determined the suspicious activity was over 900 meters from my position. I called SPC JY and ordered him to range that group of males with kids. As I continued to watch, SPC JY answered, "812."

"Roger." Now, I could see kids moving in and out of their little circle. As the kids moved, I saw two adult males digging near the base of the pole.

I said, "They are digging."

Even on the battlefield, pulling the trigger didn't sit right with me. The presence of children raised doubts, their innocence at odds with the chaos. The adult male's deliberate movements unsettled me—a subtle act that hinted at something more. Still, I could not shake the doubt. I couldn't label his intent as overtly hostile. *Was I missing something?* The unease clung to me, heavy as the dust in the air.

I told SGT RT, "Tell BN, two adult males with white head covers and white beards wearing light-colored man dresses and accompanied by seventeen young school-age children are gathered near the base of the electric pole, grid 12345678. MAMs are digging near the pole." I added, "Tell them we're unable to identify a clear threat but to push the info to the Marines."

SGT RT replied, "Roger."

I continued to watch them. Again, the kid's movement allowed me to see two males removing long rods from the ground, consistent with what appeared to be rifles.

I said to my team, "They are pulling rifles from the ground. RT, get your rifle. Two males, at the base of the pole, wearing a white headdress and tan-colored man dress, are digging and producing rifles." I added, "JY, spot for him."

I hefted my rifle into my shoulder, feeling the comforting weight as I acquired my target. They seemed almost ghostly at this distance, mere wisps of white and elongated black shapes. The children's brightly colored forms stood out like beacons, their innocence stark against the grim reality surrounding us.

I momentarily pulled back from the scope, calibrating my DOPE, adjusting the cam settings to 8 plus one and holding just above the head. The palms swayed gently in the still air, a silent testament to the absence of wind, leaving the world eerily quiet.

My eyes swept back to the tree line, sharp and vigilant, scanning for any sign of movement. The palm fronds rustled slightly from an unseen disturbance. I quickly adjusted my focus to the road ahead, where a shimmering mirage danced along the pavement and rose into the air. The area around us remained still. *Thank you, Lord.* I murmured, "No value wind."

I silently went through the protocol. *Close your eyes, open them, adjust. Close your eyes, open, breathe, aim, relax, fire.*

The first shot hit one of three men standing close together. As if on cue, the children's heads moved. They had heard the round coming in. Several children opened their circle, exposing the adults. After their first hesitation from shock, they soon dashed away from death.

The next round hit its intended target, but I could not determine where it hit. I looked for my second target. The injured man was lying on the ground, not moving, but alive. He appeared to be encouraging the adult males to run away. However, one of the men returned to his fallen comrade. As he reached down to help him, I squeezed my trigger, and the second one fell.

I loaded again and scanned my lane. Another male was running toward the palm grove. His back became a wider pale line. I pulled the trigger. "Hit," I called out. The man slapped the air and disappeared out of view.

I scanned to the left and located two more males running left to right and employed the ambush technique. I lined up on a pole. As he ran holding his man dress in his hand, I counted his movement from mil-dot to mil-dot. I moved to the next pole, lined up my shot, and let my air out. My crosshairs hovered on air as he entered my angle of death. I squeezed my trigger, and he fell from view.

I heard closer shots and watched SGT RT engage with his rifle. I put my rifle down to spot for him. A male near the pole grabbed for an object on the road. As I began to call out his actions, I heard a shot. The target's head snapped up, almost looking at the sky, and slowly he fell from view.

I told SPC JY to call it in. "Tell them five MAMs were discovered digging on the previously said grid, using children as human shields. Four of the MAMs were hit by the Joker 4 team, and the fifth male was able to escape Joker 4's efforts."

Based on our actions, I decided to move us off the roof. "Start building up fighting positions. We shot too many times. Let's assume we have been compromised." I did a quick LACE assessment (checking for Liquids, Ammunition, Casualties, Equipment) to verify our status, then began to build up a balcony on the second floor. As we neared completion, security reported, "Joker 4, we have people at the courtyard door." As I started to speak, I heard the term "friendlies."

Dismayed, I said, "Bullshit."

Security replied, "Seriously, there is a 113 outside the door."

I looked at the team. "Stay in your position. I will go outside. Prep claymores." I went outside, drew my pistol, and shouted, "Who are you guys?"

I could hear laughter. "Friendlies!" The accent sounded right and lighthearted. Curiosity got the best of me.

Slowly, I opened the courtyard door. Framed in the door stood a tall American in a desert camouflage uniform, helmet, and sunglasses. Directly behind him was the open rear (ramp down) of an American tactical 113 vehicle. He exclaimed, "Hey, we heard your team has a good position, and we want to help."

I was annoyed, but I knew who I was speaking to. "What do you mean, help?"

"Let us inside, and we want to set up."

I asked. "SEALs?"

He did not answer me, but his grin confirmed the answer.

One does not argue with SEALs. I stepped back, watched them enter the courtyard and proceed to the resident's front door. Once inside, I gave them a quick SITREP and added, "We are leaving. This house has been marked and compromised. If I can offer any advice, if you choose to remain behind, stay off the roof."

I turned to my team, "Collect your gear, check SI, and prepare to move out. We are RTB. Call in movement to BN and ECP 8. The route will be south on this street, move to ECP 8, enter lines, and move to Corregidor from ECP 8. I will take point. Set up channels to monitor all nets as we reach ECP 8."

The SEALs said little as they watched us pack up. I showed them our fighting positions and departed with the words, "Be safe."

That night, back at Camp Corregidor, the reporter and I were talking on a rooftop. He had expressed his wish to file a report with UPI and thought our shooting was terrific. "I want to write up today's incident."

I told him not to. As snipers, we were shooting at a wall of flesh. We could not miss those kills. We had eleven months in Ramadi with a year's worth of real-world engagements every week. Today was just another day. "Sir, with all due respect, it is not newsworthy." To my knowledge, it never made it past SIG ACTS (significant acts).

While hitting a moving target can be difficult, the man was frightened and moved slowly. We were so far away that they could not tell where we were shooting from. They were scared, and their bold recklessness and poor planning had cost them their lives.

CHAPTER 13

TO THE VERY END

With my sons (Joshua, James, and the Author) in Sinjar, Iraq, fighting with Yazidi Peshmerga against ISIS 2015.

The following day, I arrived at ECP 8 to provide assistance and give the other team a break. A few hours later, we heard a commotion in the street east of our position. Like a swarm of horseflies, the air filled with rounds snapping around our position on the rooftop, coupled with the scream of a rocket barking off its launcher. I fought my desire to curl into a fetal position and cover my head. I went to a kneeling position, toward a line of Iraqi soldiers defending a wall to the east.

In the corner under the overhead cover, one Iraqi Army soldier held his PKM machine gun, officially known as the Pulemyot Kalashnikova 7.62 54mm general-purpose machine gun (MG). He fired haphazardly in the direction of the enemy while crouching behind the wall, below the MG. His hand reached up to the trigger, squeezing burst after burst—a totally useless action in a firefight.

I ran towards him and grabbed the PKM. For a moment, I felt awkward. *How do I fire from this stock?* Getting past my lack of training on foreign weapons, I gained positive control and engaged the insurgents.

I was grateful for the overhead cover. When the MG went silent, I began to load it. As I lifted the cover and turned my face away, an IA soldier ran towards me under fire across the roof. He carried my helmet. I immediately grabbed the helmet, put it on, and fastened the chin piece. As I resumed fire, the IA Soldier said, "Please, Martinez, wear your helmet. You can die out here."

He headed back into an exposed position to check on his IA soldiers. I stopped shooting to determine targets and noticed a vehicle on the south side of the street. A man was leaning through a vehicle window trying to grab something inside.

AQI peeled away from the fight. I was curious to see what that man wanted from his vehicle. I grabbed a team and proceeded toward the vehicle. Understandably, no one cared, and no one volunteered.

Nearby, I heard a voice in broken English say, "I will go with you." With a hand motion slapping his ass, he added, "I will cover you."

Almost laughing, I agreed and composed a quick plan. Get to the vehicle, do a cursory search, and return to ECP 8. My team would provide the overwatch.

We exited the house and crossed the street on the north side. The IA Soldier followed me. When we reached within twenty feet of the vehicle, I decided I would take a look and destroy the vehicle with a hand grenade. Crouched and moving with my weapon at the high ready, I moved across the street. I glanced into the buildings I passed.

When I reached the vehicle from the passenger side, I saw nothing that grabbed my attention. Feeling vulnerable and exposed, I yanked a hand grenade from my vest, prepped it, and threw it into the front seat. Turning, I realized I was too close to the potential blast. I literally took three steps and threw myself behind a mound of dirt covered in broken bricks and exposed rebar. As I landed, I waited for the pain—confirming this idea was stupid.

After the explosion, I looked at the IA soldier who was grinning from ear to ear. After I had no injury, I ran to my IA soldier. "Let's go, follow me."

Upon returning to ECP 8, I found myself pondering life here in this city. As the sun dipped below the horizon, I stared at the distant city lights and felt the faint hum of life beyond our world. Somewhere out there, children were laughing. Families were gathering for dinner. Here, the only warmth came from the heat of the rifle barrel and the cold fire of vengeance. I reached for the anger and let it settle in my chest like armor. As quickly as I grabbed it, I forcefully tucked it into a corner of my mind. Anger can make one impulsive. I needed to be cool and calm, sharp and focused, feeling alive. While this worked, it was stripped away my human emotions.

The fight was coming. We knew it as surely as we knew the sound of our own breath. I tightened my grip on the rifle, squinting into the endless expanse of night, waiting for the inevitable flash of light or crack of gunfire that would signal it had begun again.

In a flash, I understood the truth I had been avoiding. The war wasn't only fought on these streets. It was fought within us, and it was winning.

"The war wasn't only fought on these streets. It was fought within us, and it was winning."

We needed to finish construction on ECP 8, so teams performed overwatch for the Soldiers working on the site. Quietness did not exist on those four corners. The enemy also came from the north through a small palm grove and used buildings to disrupt and destabilize our efforts to complete the project.

I knelt against the wall, my back pressed into the rough surface, facing north toward the fighting. The dry, acrid air was thick with the stench of gunpowder and sweat. From the corner of my eye, I caught movement—men spilling from the second floor of a crumbling building, their figures darting in desperation as they made for the cover of the palm grove.

Here they come.

My hands steadied my M16A4 as I raised it, aligning my optic. The world narrowed to the scope. Adrenaline sharpened every detail. One man appeared at the top of the staircase outside the building, his steps quick, his intent clear. My breath slowed as training took over. Finger on the trigger, I waited. The moment stretched impossibly long as he descended the stairs, each step bringing him closer into my sights.

I squeezed.

The crack of the rifle was deafening, momentarily overwhelming the cacophony around me. Through the optic, I saw him falter, his hand shooting to his abdomen as his body twisted with the force of the impact. His face contorted in agony before he crumpled, tumbling down the staircase like a broken marionette. He hit the ground hard, writhing in pain, his struggle bleak against the unforgiving terrain.

No pause. No time to process. The brutal rhythm of war pressed on and on, the noise a constant roar in our ears. This was the Global War on Terrorism in 2006, and mercy was a flower that did not grow in our garden.

As I sighted in on my next target, a force slammed into me, lifting me like a rag doll and slamming me to the ground. When I opened my eyes, the world was oddly silent.

Above me was serene blue sky. So beautiful. It felt surreal. The silence was profound, almost holy. *Was I dead?* This must be the transition, the moment God shows you the other side.

Then, training kicked in. No one was coming. If I wanted to live, I had to move. My trembling hands reached for my helmet, sliding it off with an almost dreamlike detachment as I groped for the wound. Nothing. No blood, no mortal injury—just hands caked with the filth of Ramadi: dirt, dust, grime, and the residue of battle.

I was alive.

The realization hit me as the noise trickled back in, faint and distant at first, then crashing over me like a wave. The familiar staccato of rifles firing, of shouted commands, and the relentless fight of my brothers-in-arms. I pushed myself upright, disoriented and sluggish, my body protesting as though time itself were dragging its feet.

Where is my weapon? My link to the fight had vanished. I looked toward the wall where 506 scouts and snipers held their ground with unyielding ferocity, delivering precision fire that could only be described as pure, concentrated hate. They fought with a ruthlessness born of desperation and loyalty, their actions a testament to the bond that held us together in this trial of violence.

I wanted to be with them, to rejoin the fight and shoulder my part of the burden. My body moved too slowly, too awkwardly, as if held down by a powerful force. By the time I managed to crawl forward, the fight was over.

From the wall, my brothers reloaded, redistributed ammunition, and called out names for accountability, their voices sharp and urgent. The lull after the storm was thick with exhaustion and relief, a fragile moment where life and death stood side by side. I sat there, alive, shaken, and grateful, staring at the dirt on my hands and the sky above me—both grim and beautiful reminders of the narrow line we tread in a state of war.

I stayed quiet, still feeling removed from the scene. I saw WD smiling, SPC JY reloading his weapon, and Bennett repositioning his body for the next fight. Team leaders were moving down the line, checking on their Soldiers. Later, when I checked my helmet, I expected to see a crease or actual penetration, but my helmet was clean of any such indicators. To this day, I have no idea what happened to me on that roof.

Going Home

Returning to Camp Corregidor felt unreal, as if I were caught between worlds. My time in Ramadi was drawing to a close, but the endless and unrelenting war would grind on. When ordered to report to the Animal Company Commander, I entered his quarters without hesitation, carrying with me the weight of battles fought and the inevitable reality of leaving them behind.

The CO greeted me near his cot, his expression subdued and his tone even. He motioned for me to sit in a small black range chair nearby. Dressed in shorts and a T-shirt, barefoot, he looked utterly drained, the exhaustion of command lined into his face. His burdens were immense, more than the logistical challenges of war. They were deeply personal. His Soldiers were not numbers in a roster. They were men from the same state, the same communities, and the same schools. The responsibility of leading them, protecting them, and sometimes watching them come back injured had carved into his very posture.

For a moment, neither of us spoke. The silence felt heavy with unspoken truths. When he spoke, his words carried the same weight as his presence: resigned, steady, and weary. The surreal nature of this war and my place in it settled over me like a quiet surrender. The battles would continue, lives would be lost, and leaders like him would bear the impossible load long after I was gone.

"You've done an incredible job here. Many argued you had gone rogue, but I defended you. Your lane in this fight was always different, and it's hard to argue against someone whose actions breed success. You trained your team. You planned with them, led them through a difficult year, and now you've completed your final mission. Well done, SSG Martinez. After you turn in your equipment, Joker 4 will be on the first chalk out of here. Dismissed."

I stood up, snapped to attention, executed an about face, and walked out. The moment was bittersweet. It was over. For the next few days, I endured the monotonous tasks of turning in gear and watching the 506 continue their fight. Part of me longed to be out there with them. Deep in my heart, I carried a belief that I had not done enough. I had not killed enough insurgents who had brought so much death and suffering to this land.

I briefly considered extending my deployment and staying with the 506. When I floated the idea to my (now-ex) wife, her meltdown on the phone was enough to silence the thought. I hung up. I had no energy to argue.

With my gear packed and my weapons turned in, I went down to the first floor, sat with my back to the wall, and waited, detached from the moment. Finally, word came move to the helo pad. The familiar silhouette of two CH-46 helicopters emerged in the darkness, blackout conditions masking the noise of the night. I moved up the ramp, my body on autopilot, my mind elsewhere. The bird lifted into black skies, and twenty minutes later, we landed in TQ.

It was over. Or so I thought.

Sitting outside that night, under the vast, silent sky, it hit me. The war was far from over. Another city, another culture, another mission—they were all inevitable. The war would rage on for years, but I was leaving it behind. My brothers would continue fighting, enduring the same brutality we had all faced. I told myself I had done enough, but a quiet voice inside me insisted otherwise. I should have stayed with my brothers, departing only when the 506 left the theater.

The weight of war followed me home. It never truly leaves a Soldier. The echoes of Ramadi would creep into our homes, haunt our memories, and shadow our relationships. We had walked with Death for so long, it became a permanent stamp on our minds that would stay until the end of our days.

Victory was a fragile concept, defined not by territory gained or battles won, but simply by coming home in one piece. We struggled to redefine joy, to find hope in small moments, to forge new journeys, and discover peace. We passed the torch to a new generation, hoping they would find something we had not, a better reason, a brighter ending, or at least a quieter burden.

During the war and the tours that would follow after Ramadi, I accepted that I was not a religious man, nor did I place much weight on the word of God. Southwest Asia was a land consumed by war and surrounded by atrocities carried out in His name. It was a boiling kettle of horrors, each act cloaked in the banner of divine justification. Now, my relationship with God is through my quiet time and daily prayers—a personal expression of belief that dismisses the relevance of any religion. I am grateful I found Him again.

War is a dark alchemist, transforming the very essence of a man. In that pressure, I searched desperately for meaning—for justice, for a sense of righteousness that might act as a shield against the carnage. Instead, I found confusion, pettiness, lies, and anguish. The truths I sought unraveled into shadows. The boy inside me burned away in that inferno. In his place emerged a man, changed, hardened, and forever marked by what he had witnessed—the human cost of being a warrior.

In coming home, I lived with sadness tempered by acceptance, loss tempered by hope, and longing for something simpler with the fragile promise of healing and rebuilding. A Soldier's farewell is not about leaving the battlefield. It is learning how to live again.

AUTHORS NOTE

Department of Defense Official Photo SGM Harry Martinez, _date_

My name is Sergeant Major Harry Martinez. What you have read was written to lay bare the reality of war in the lives of citizen Soldiers swept into its path. Ordinary men found themselves thrust into extraordinary circumstances, captured by honest conversations, documented accounts, and memories seared by fire and steel. The dialogue here is a blend of recollections reconstructed from interviews, stories, and official reports, each detail a testament to the vicious truth of combat.

This book is not a statement on the morality of war, nor is it a plea for peace. I am under no illusions about the nature of humanity. War is in our blood, and violence is an enduring legacy. I am both a witness and a willing participant in that darkness, marked by it, and forever changed. I chose to fight. I volunteered to step into the heart of savagery and violence, not once but again and again. I volunteered to join the Army National Guard. I volunteered for Combat Arms, and I volunteered to become a sniper. I did two more combat tours. I served five months in northern Iraq, holding the line against ISIS near the Syrian border as a Peshmerga Yazidi fighter, this time with my two sons, who are former Marines… but that is another story for another time.

For better or worse, this is the reality of war through the eyes of those who lived it.

ABOUT THE AUTHOR

SGM (Ret.) Harry R. Martinez 2026

Harry R. Martinez is a former enlisted Marine and U.S. Army infantryman with more than nine years of active-duty service, followed by continued service in the Army National Guard, from which he retired at the rank of Sergeant Major. His service record includes participation in three conflicts in Central America and Southwest Asia, encompassing six combat deployments and two operational rotations.

He pursued academic studies in American history at American Military University. In civilian service, he served nearly fifteen years with the Princeton Township Police Department, New Jersey, where he was assigned to the Patrol Division and Emergency Response Team. He currently resides in southeastern North Carolina with his immediate family.

GLOSSARY OF MILITARY TERMS

ACU: Army Combat Uniform

AD: Active Duty

Ali Baba: term used by American and CF units to designate "bad guy"

ANGLICO: USMC unit, Air Naval Gunfire Liaison Company

AO: Area of Operations

AQI: Al-Qaeda Iraq

AR: Amor

ARNG: Army National Guard

Battle Captain: Officer in charge of TOC

BDU: Battle Dress Uniform

BN: Battalion

BTHP: Boat Tail Hallow Point

BFV: Bradley Fighting Vehicle

CH 46- 47: Twin engine tandem rotor heavy helicopter for troop and equipment transport

Charlie Mike: Continue the Mission

Coalition Forces: Coalition Forces

Co: Company

COP: Combat Out Post

DCU: Desert Combat uniform

ECP: Entry Control Point

EOD: Explosive Ordinance Disposal

ERT: Emergency Response Team

FBI: Federal Bureau of Investigation

FMF: Fleet Marine Force

FMJ: Full Metal Jacket

FOB: Forward Operating Base

FRAGO: Fragmentary Order

GWOT: Global War on Terrorism

Hajji: disparaging slang term among American and CF units for indigenous personnel

HE: High Explosive

HMMWV: High Mobility Multipurpose Vehicle, pronounced "HUMVEE"

HQ: Headquarters

IA: Iraqi Army

IBCT: Infantry Brigade Combat Team

ICOM: handheld radio used by the sniper section manufactured by International Communication Company

IED: Improvised Explosive Device

IOTV: Improved Outer Tactical Vest

Klick: Slang for Kilometer

LAPD: Los Angeles Police Department

LMAV: Laser Maverick Missile

LRSD: Long Range Surveillance Detachment

M24: In GWOT, up to 2010 for conventional forces, was the Army's sniper weapon system based on the Remington 700 platform primarily chambered for the BTHP 7.62 x51mm (308 match grade round). Using various optics. Effective range 800 meters. However, can exceed the established range, shooter dependent

M113: Lightweight armored personnel carrier

M203: Single shot 40 mm grenade launcher deigned to be attached to the M16 and M4 series

M240B: A machine gun chambered primarily for the 7.62 caliber weighing approximately 27.6 lbs. with an effective range up to 1200 meters

M249: is a light Machine gun chambered for the NATO 5.56 weighing approximately 17 lbs. with an effective range up to 800 meters.

Magazine: Designed to hold ammunition and rounds into a chamber, pistol and rifle

MAM: Military Age Male

MAR DIV: Marine Division

Mech: Mechanical

MG: Machine Gun

MOPP: Mission Oriented Protective Posture against chemical, nuclear, biological and radiological threats

MSR: Main Supply Route

MWR: Morale Welfare Recreation Center

NCO: Non-Commission Officer, i.e. SGT through SGM

NVGs/NODs: Night Vision Goggles/ Night Observation Device

OOM: Order of March

OP: Observation Post

OPS: Operations

PCI/PCC: Pre-combat inspection (Soldier)/ Pre-combat check (equipment)

PID: Positive Identification

PKM Machine Gun: Russian Pulemyot Kalashnikova Modernizironvanny 7.62x54 caliber heavy machine gun

PL: Army Platoon Leader (officer)

PLT: Platoon

PSG: Platoon Sergeant, title/position in a platoon, E7

PT: Physical Training

QRF: Quick Reaction Force

RAVEN: Drone

RC: Reserve Component

ROE: Rules of Engagement

RPG: Rocket Propelled Grenade

RTB: Return to Base

RTO: (Old Term) Radio Telephone Operator

S1: Staff administration

S2: Staff Intelligence section

S3: Staff Operations section

S4: Staff logistics/Supply

SAF: Small Arms Fire

SAPI: Small Arms Protective inserts (armor)

SDM: Squad Designated Marksman

SEAL team: Sea Air Land team

SF: Army Special Forces

SHADOW 4: Call sign and numerical value indicate respective team

SIGACTS: Significant Activities

SIGINT: Signal Intelligence

SINCGARS: Single Channel Ground and Airborne Radio System

SITREP: Situation Report

Snivel gear: Slang for cold weather gear

SQD: Squad

SSOC: Scout Sniper Ops Center

STA: Marine Scout Sniper PLT for Surveillance Target Acquisition

SWAT: Special Weapons and Tactics

Title 10: Code for Federal Funding and Active Duty

TF: Task Force

TOC: Tactical operations center

TOP: Slang for 1SG

TTP: Tactics Technique and Procedure

TQ: Marine Base Taddqum Iraq Anbar Province

ARMY Rank/Grade (Pay)

PVT (no stripe): E1

PVT (one stripe): Private E2

PFC: Private First Class E3

SPC: Specialist E4

CPL: Corporal E4

SGT: Sergeant E5

SSG: Staff Sergeant E6

SFC: Sergeant First Class E7

MSG: Master Sergeant E8

1SG: First Sergeant E8

SGM: Sergeant Major E9

CSM: Command Sergeant Major E9

LT: Lieutenant Officer O/1/O2

CPT: Captain O3

MAJ: Major O4

LTC: Lieutenant Colonel O5

COL: Colonel O6

USMC: United States Marine Corps

VBIED: Vehicle Borne Improvised Explosive Device, pronounced V-bed

XO: Executive Officer

X- Ray: Phonetic spelling for letter "X" also used to designate the BN TOC

www.ingramcontent.com/pod-product-compliance
Lightning Source LLC
LaVergne TN
LVHW052250200926
848481LV00002B/324
* 9 7 8 1 9 7 2 0 5 9 0 0 5 *